NUTRITIONAL DATA
FOR
UNITED STATES
AND
CANADIAN FEEDS
Third Revision

United States–Canadian Tables of Feed Composition

Subcommittee on Feed Composition
Committee on Animal Nutrition
Board on Agriculture and Renewable Resources
Commission on Natural Resources
National Research Council

NATIONAL ACADEMY PRESS
Washington, D.C.   1982

NOTICE: The project that is the subject of this report was approved by the Governing Board of the National Research Council, whose members are drawn from the councils of the National Academy of Sciences, the National Academy of Engineering, and the Institute of Medicine. The members of the committee responsible for the report were chosen for their special competences and with regard for appropriate balance.

This report has been reviewed by a group other than the authors according to procedures approved by a Report Review Committee consisting of members of the National Academy of Sciences, the National Academy of Engineering, and the Institute of Medicine.

The National Research Council was established by the National Academy of Sciences in 1916 to associate the broad community of science and technology with the Academy's purposes of furthering knowledge and of advising the federal government. The Council operates in accordance with general policies determined by the Academy under the authority of its congressional charter of 1863, which establishes the Academy as a private, nonprofit, self-governing membership corporation. The Council has become the principal operating agency of both the National Academy of Sciences and the National Academy of Engineering in the conduct of their services to the government, the public, and the scientific and engineering communities. It is administered jointly by both Academies and the Institute of Medicine. The National Academy of Engineering and the Institute of Medicine were established in 1964 and 1970, respectively, under the charter of the National Academy of Sciences.

This study was supported by Agricultural Research, Science and Education Administration of the U.S. Department of Agriculture; by the Bureau of Veterinary Medicine, Food and Drug Administration of the U.S. Department of Health and Human Services; and by Agriculture Canada.

Library of Congress Cataloging in Publication Data
Main entry under title:

United States—Canadian tables of feed composition.

Bibliography: p.
1. Feeds—Composition—Tables. 2. Feeds—United States—Composition—
Tables. 3. Feeds—Canada—Composition—Tables. I. National Research Council
(U.S.). Subcommittee on Feed Composition.
SF97.U56 1982   636.08'55   82-3625
ISBN 0-309-03245-8   AACR2

Available from

NATIONAL ACADEMY PRESS
2101 Constitution Avenue, N.W.
Washington, D.C. 20418

Printed in the United States of America

PREFACE

This report is the third revision of the joint United States–Canadian Tables of Feed Composition, NAS-NRC publication 659, issued in 1959. The first revision, publication 1232 (1964), consolidated the data in the first joint report with selected data from NAS-NRC publication 449 and NAS-NRC publication 585. The second revision was NAS-NRC publication 1684 (1969).

The feeds included here were selected by the Subcommittee on Feed Composition and approved by the Committee on Animal Nutrition and its subcommittees on nutrient requirements of domestic animals. This report brings together analytical data on more than 600 feeds. Data are presented on 68 attributes (nutrients). The report provides working tables for feed manufacturers, nutritional research scientists, teachers, students, county agents, and farmers to use as adjuncts to reports in the NRC nutrient requirement series.

This study was partly supported by financial assistance to Utah State University from the U.S. Department of Agriculture (USDA) and from the Agricultural Experiment Station, Utah State University. Support for subcommittee activities was received from Agricultural Research, Science and Education Administration, USDA; the Bureau of Veterinary Medicine, Food and Drug Administration, U.S. Department of Health and Human Services; and Agriculture Canada, Ottawa, Ontario.

The subcommittee wishes to thank the many scientists in commercial and university laboratories who supplied data that have been used in compiling the information contained in this report. We are grateful to the Technical Committee of the USDA Cooperative Regional Project S-45 for providing data on a number of forages grown in the southeastern section of the United States. Special thanks are due L. C. Kearl, P. V. Fonnesbeck, and Howard Lloyd of the International Feedstuffs Institute, Utah State University, for their untiring efforts and special competencies in compiling and organizing the data. We are indebted to Philip Ross and Selma P. Baron of the Board on Agriculture and Renewable Resources for their assistance in the production of this report and to the members of the Committee on Animal Nutrition for their critical reviews and suggestions.

We want to extend our special thanks to Donald L. Bath, Carl E. Coppock, Eugene S. Erwin, Steve Leeson, Fredric N. Owens, John V. Shutze, Milton L. Sunde, and Eric W. Swanson who reviewed the draft of the report and made helpful comments and suggestions for our consideration. The report was also reviewed by Bernard S. Schweigert for the Board on Agriculture and Renewable Resources and by Howard S. Teague for the Commission on Natural Resources.

Subcommittee on Feed Composition

JOSEPH H. CONRAD, *Chairman*, University of Florida

CHARLES W. DEYOE, Kansas State University

LORIN E. HARRIS, Utah State University

PAUL W. MOE, USDA, Beltsville, Maryland

RODNEY L. PRESTON, Texas Tech University

PETER J. VAN SOEST, Cornell University

CONTENTS

INTRODUCTION

The need for information concerning the nutritive value of feeds was recognized long ago. The first tables (ca. 1800–1810) were based on the relative amounts of feed required to maintain and support animal production (Tyler, 1975). Later in the nineteenth century, German and French scientists developed the crude fiber analysis procedure and partitioned feeds into nitrogenous and carbohydrate fractions. The work continued with the early digestion trials and the appearance of tables containing digestibility and proximate analysis (Henneberg and Stohmann, 1860, 1864). Scientists in the United States expanded on the Europeans' work and published feed tables containing nutrient and energy values (Atwater, 1874; Henry, 1898; Armsby, 1903; Henry and Morrison, 1910; and Morrison et al., 1936).

The need for a review of feed composition information was recognized by the National Academy of Sciences in 1952.

This resulted in a publication on the composition of concentrates (National Academy of Sciences, 1956) and one on the composition of forages and grains (National Academy of Sciences, 1958). A number of comprehensive tables of feed composition, including feeds representing different geographical areas, have been published during the last decade. These include the following: *Atlas of Nutritional Data on United States and Canadian Feeds* (National Academy of Sciences, 1971); *Applied Animal Nutrition* (Crampton and Harris, 1969); *Latin American Tables of Feed Composition* (McDowell et al., 1974); *Tropical Feeds* (Göhl, 1975); *Nutrient Composition of Some Philippine Feedstuffs* (Castillo and Gerpacio, 1976), *Composition of British Feedstuffs* (Agricultural Research Council, 1976), and *Middle East Feed Composition Tables* (Kearl et al., 1979).

COMPOSITION OF FEEDS

Tables 1–6 present the composition of important United States–Canadian feeds. Nutrient concentrations are organized as follows:

Table 1 Energy values, proximate analyses, plant cell wall constituents, and acid detergent fiber
Table 2 Mineral composition
Table 3 Vitamin composition
Table 4 Amino acid values
Table 5 Fat and fatty acid values
Table 6 Mineral supplement composition

INTERNATIONAL FEED NOMENCLATURE

The nomenclature of the feeds under which the analytical data are shown primarily follows the International Feed Vocabulary of Harris et al. (1980, 1981). Many feeds in the United States have official names and definitions designated by the Association of American Feed Control Officials (AAFCO, 1979). Frequently, however, these names are common or trade names and the origin of the feed name does not follow a standardized naming system.

The International Feed Vocabulary is designed to give a comprehensive name to each feed as concisely as possible. Each feed name was coined by using descriptors taken from one or more of six facets:

1. Origin consisting of scientific name (genus, species, variety) and common name (generic name; breed or kind; strain or chemical formula)
2. Part fed to animals as affected by process(es)

3. Process(es) and treatment(s) to which the part has been subjected
4. Stage of maturity or development
5. Cutting (applicable to forages)
6. Grade (official grades with guarantees)

See Table 7 for stage of maturity terms for plants.

A complete International Feed Name consists of all descriptors applicable to the feeds. Definitions for the part and process descriptors are given by Harris et al. (1981).

INTERNATIONAL FEED CLASSES

Feeds are grouped into eight classes on the basis of their composition and their use in formulating diets (Table 8). These classes, by necessity, are arbitrary, and in borderline cases a feed is assigned to a class according to its most common use in typical feeding practice.

INTERNATIONAL FEED NUMBER (IFN)

Each International Feed Name is assigned a five-digit International Feed Number (IFN) for its identification. This numerical representation is the link between the International Feed Names and chemical and biological data in the USA databank. The numbers are particularly useful as a tag to recall the nutrient data for calculating diets. The Feed Class Number (Table 8) is entered in front of the IFN when feed tables are prepared.

ANALYTICAL AND BIOLOGICAL DATA

SOURCE OF DATA

Most of the data was compiled by the International Feedstuffs Institute at Utah State University, Logan, Utah. However, data from many individuals in both industry and public institutions have been incorporated.

To assist in making the tables more useful, source data values were generated for missing data for some attributes by using regression equations as outlined below. In some cases, such as for stage of maturity of forages, data were estimated from similar feeds. When reasonable values could not be estimated or were insignificant in formulating animal diets, the spaces were left blank.

Data in this report may differ from those in various other NRC reports because of the reasons given above, but the values in the tables represent the best judgment of the Committee on Animal Nutrition's Subcommittee on Feed Composition.

VARIATION IN DATA

Feedstuffs are not of constant composition, and individual feed samples may vary widely from the values set forth in these tables. The variation is caused by such factors as variety, climate, soil, and length of storage. Actual analysis should be obtained and used wherever possible. Often, however, it is either impossible to determine actual composition or there is insufficient time to obtain such analysis, making tabulated data the next best source of information.

When tabulated data are used, it should be understood that feeds do vary in their composition and, therefore, the values should be used as guides. Organic constituents (e.g., crude protein, cell wall constituents, ether extract, amino acids) can vary as much as ± 15 percent, the inorganic constituents as much as ± 30 percent, and the energy values as much as ± 10 percent.

See Table 9 for weight-unit conversion factors.

DRY MATTER

Typical dry matter values are shown; however, the moisture content of feeds varies greatly and the dry matter content may be the main reason for variation in the composition of feedstuffs on an "as-fed" basis. Because dry matter can vary greatly and because one of the factors regulating total feed intake is the dry matter content of feeds, diet formulation on a dry matter basis is preferred over using the as-fed basis. Dry matter values of nutrient attributes may be converted to an as-fed basis by simply multiplying the dry matter by the nutrient values and dividing by 100.

ENERGY VALUE OF FEEDS

Energy values of feeds are frequently influenced by interactions with other feeds, by level of feed intake, and by other management factors. The values listed in this publication are, therefore, a guide in "normal" feeding and management situations and should not be considered to be an inflexible constant.

Because of the effect of level of intake on digestibility of feeds, the total digestible nutrients (TDN), digestible energy (DE), and metabolizable energy (ME) values of feeds for ruminants have been listed as appropriate for animals in production.

Energy values for ruminants, horses, and swine include TDN, DE, and ME. For ruminants, net energy values are given for maintenance (NE_m), gain (NE_g), and lactation (NE_l). For poultry, energy values include nitrogen-corrected metabolizable energy (ME_n), true metabolizable energy (TME), and net energy for production (NE_p). A discussion of these net energy values may be found in individual NRC nutrient requirement reports. Details of methods of calculating individual energy values are as follows (all calculations are done on the dry matter basis):

3

Energy Values of Feeds for Ruminants

Total Digestible Nutrients Total digestible nutrients for ruminants was calculated from:

a. average TDN for cattle and sheep
b. or from digestion coefficients for cattle and sheep as:

digestible protein (%)	× 1.0
digestible crude fiber (%)	× 1.0
digestible nitrogen-free extract (%)	× 1.0
digestible ether extract (%)	× 2.25
TDN (%)	TOTAL

Digestible Energy Digestible energy for cattle and/or sheep was calculated by using the formulas of Crampton et al. (1957) and Swift (1957):

$$\text{DE (Mcal/kg DM)} = \text{TDN for cattle and sheep} \times 0.04409.$$

Metabolizable Energy Metabolizable energy was calculated from DE. These values were used to calculate NE_m and NE_g from the following formula:

$$\text{ME (Mcal/kg DM)} = 0.82 \times \text{DE (Mcal/kg DM)}$$

For the ME shown in the table, the following formula was used (Moe and Tyrrell, 1976; NRC, 1981):

$$\text{ME (Mcal/kg DM)} = -0.45 + 1.01 \text{ DE (Mcal/kg DM)}$$

Net Energy Net energy for finishing cattle was calculated by equations developed by Garrett (1977):

$$NE_m \text{ (Mcal/kg DM)} = 1.115 - 0.8971 \text{ ME} + 0.6507 \text{ ME}^2 \\ - 0.1028 \text{ ME}^3 + 0.005725 \text{ ME}^4$$
$$NE_g \text{ (Mcal/kg DM)} = 3.178 \text{ ME} - 0.8646 \text{ ME}^2 \\ + 0.1275 \text{ ME}^3 - 0.00678 \text{ ME}^4 \\ - 3.325$$

Net energy values for NE_l were calculated by using the formula of Moe and Tyrrell (1976):

$$NE_l \text{ (Mcal/kg DM)} = -0.12 + 0.0245 \text{ TDN (\% of DM).}$$

Energy Values of Feeds for Horses and Swine

Total Digestible Nutrients Total digestible nutrients for horses and swine were calculated from:

a. average TDN
b. or from digestion coefficients as:

digestible protein (%)	× 1.0
digestible crude fiber (%)	× 1.0
digestible nitrogen-free extract (%)	× 1.0
digestible ether extract (%)	× 2.25
TDN (%)	TOTAL

c. DE for horses (Fonnesbeck et al., 1967; Fonnesbeck, 1968):

$$\text{TDN \%} = 20.35 \times \text{DE (Mcal/kg)} + 8.90$$

(This formula was used only for class 1 feeds.)
d. TDN for horses and swine was not calculated from ME
e. or from regression equations (Harris et al., 1972).

Digestible Energy Digestible energy for horses and swine was calculated from:

a. the average digestible energy in kcal/kg or Mcal/kg
b. DE (kcal/kg DM) = Gross Energy (kcal/kg) × Gross Energy digestion coefficient
c. TDN for horses (Fonnesbeck et al., 1967; Fonnesbeck, 1968):

$$\text{DE (Mcal/kg DM)} = 0.0365 \times \text{TDN \%} + 0.172$$

d. TDN for swine (Crampton et al., 1957; Swift, 1957)

$$\text{DE (kcal/kg DM)} = \text{TDN \%} \times 44.09$$

Metabolizable Energy Metabolizable energy for horses, swine, and poultry was calculated from:

a. the average metabolizable energy in kcal/kg or Mcal/kg
b. the average true metabolizable energy (TME) in kcal/kg for poultry (Sibbald, 1977)
c. the average nitrogen corrected metabolizable energy (ME_n) for poultry (National Research Council, 1966)
d. ME for horses (Mcal/kg DM) = 0.82 × DE (Mcal/kg DM)
e. ME for swine (Asplund and Harris, 1969)
$$\text{ME (kcal/kg DM)} = 0.96 - (0.00202 \times \text{crude protein \%}) \\ \times \text{DE (kcal/kg DM)}$$

PROTEIN

Crude Protein The crude protein value shown in these tables is the nitrogen value times 100/16 or 6.25, because protein on the average contains 16 percent nitrogen. To determine the apparent protein content of a given feed more accurately, conversion factors for that feed can be used; however, these factors have been determined for only a few feeds (Jones, 1941). Crude protein values do not distinguish between true protein and nonprotein nitrogen content of feeds.

Digestible Protein Digestible protein was not included in Table 1 but it can be calculated for each kind of animal as follows:

a. Digestible protein = $\dfrac{\% \text{ crude protein} \times \text{protein digestion coefficient}}{100}$

or

b. By equations developed for six animal species and four feed classes by Knight and Harris (1966).

Because of the large contribution of body protein to the apparent protein in feces (metabolic fecal protein), the digestible protein value for a given feed can be misleading (Preston, 1972). The digestible protein content for the total diet can be more accurately calculated from the crude protein content of the diet using the equations of Knight and Harris (1966).

PLANT CELL WALL CONSTITUENTS INCLUDING CRUDE FIBER

Total insoluble dietary fiber is represented by cellulose, hemicellulose, and lignin. Plant cell walls also contain pectins, which are largely removed with neutral detergent, and protein and mineral components. Some protein fractions are very insoluble and are the slowest digesting nitrogen fraction of forages (Pichard and Van Soest, 1977).

Plant cell wall analysis quantitatively includes the truly indigestible lignified portion of the feed and is, therefore, the theoretical replacement for crude fiber. But while cell wall content is the best predictor for digestibility in nonruminants (Henry, 1976), it is more clearly related to intake in ruminants than to digestibility (Table 10). Acid detergent fiber and lignin are better indicators of digestibility for ruminant diets.

Forages were once defined as feeds containing more than 18 percent crude fiber. But it is recommended that the use of crude fiber as a means of classification be abandoned in favor of using the percentage of cell wall constituents. Hence, forages are defined as leaf and stem portions of plants with more than 35 percent cell wall constituents in the dry matter. In addition, forages can be further characterized by the percentages of their cell wall components (Table 11), which is the recommended basis for a new hay grading system (Rohweder et al., 1978).

This system recognizes that forages vary in composition according to conditions of growth: Those forages growing in warmer and wetter climates tend to be higher in lignin content. Warm-season grasses are also higher in cell wall content and often lower in protein at comparable stages of growth than are cool-season or northern grasses. Northern alfalfa tends to have lower lignin and protein and higher cell wall content than southern alfalfa. Most forages grown in cool conditions tend to be more digestible. In temperate regions, digestibility of pasture in the spring and summer is lowest at the hottest period. Autumn and forage maturity are often associated with an increase in nutritive value (Van Soest et al., 1978).

An attempt has been made to recognize the environmental and regional variables affecting forage composition (Table 1). Comparative data are given for alfalfa, Bahiagrass, Bermudagrass, fescue, pangolagrass, and sorghum.

Cellulose The most often considered carbohydrate of the fiber fraction is cellulose, which is a 1,4-β-glucan. It is the most insoluble fraction of the cell wall and is seldom obtained pure even in chemical isolation. Most celluloses contain cuticular fractions and about 15 percent arabinoxylan,

properly a hemicellulae. Most values for cellulose have been determined by the Crampton method (Crampton and Maynard, 1938) or the permanganate procedure of Van Soest and Wine (1968), which are assumed to be interchangeable.

Hemicellulose The noncellulose portion of cell wall carbohydrate is a complex substance containing a variety of linkages and sugars. One main fraction in grasses and legumes is an arabinoxylan with some glucuronic acid. Hemicellulose is not a uniform fraction and is combined with lignin in the encrusting matrix of the recovered part of the cell wall. The percentage of hemicellulose is much greater in grasses than in legumes. Values in the tables have been estimated separately for cell wall and acid detergent fiber contents.

Lignin The main organic noncarbohydrate portion of cell wall is crude lignin, composed of true lignin, cutin, Maillard polymers, and amino-protein complexes. True lignin is a phenylpropanoid polymer that provides the crosslinked three-dimensional structure that gives the plant cell wall its rigidity and resistance. It is the primary factor that reduces the digestibility of forages, although there are other contributors, such as silica in rice straw and hulls. The cuticular fraction is also resistant to digestion; it occurs in the skin surface of plants and in barks and seed hulls. It is a polymerized lipid of different constituents than lignin.

The Maillard polymer is formed upon heating and drying as the result of heat damage. It has the properties of lignin and is formed from a one-to-one condensation of amino acid from protein and a sugar unit from hemicellulose. It is indigestible and accounts for the lower protein digestibility of heated feeds. This aspect of quality is not shown in the tables. Heating in silages, hays, and pelleted feeds is highly variable and it is recommended that the availability of feed nitrogen be assayed by means of acid detergent fiber insoluble nitrogen (Goering et al., 1972).

PROXIMATE ANALYSIS AND CRUDE FIBER

The old system of feed analysis was the proximate system in which the dry matter is divided into ether extract (lipid) protein, ash, crude fiber, and nitrogen-free extract (NFE), the NFE content being determined by subtraction of the others. The principal problem of this system is the distribution of the organic nonlipid, nonprotein fraction between crude fiber and NFE, which fails to provide a meaningful separation of the carbohydrates according to their nutritive value. Crude fiber analysis fails to recover any one of the cell wall components. About 50–90 percent of the lignin, 85 percent of the hemicellulose, and 20–50 percent of the cellulose are dissolved in the determination of crude fiber content. These then are included in the calculated NFE, which is supposed to represent the available and easily digestible carbohydrates of the feed.

In about 30 percent of the analyses, NFE is determined to be less digestible than the crude fiber, primarily because most of the very indigestible lignin is included in the NFE. Because of this inaccuracy nitrogen-free extract is not reported in the tables and its use as a determinant of nutritive value is discouraged. Calculation of NFE from acid detergent fiber (lignocellulose) or cell wall content is also discouraged, because this also perpetuates inaccuracies in regard to the NFE calculated by difference.

ETHER EXTRACT

The lipid portion of plants, which is included in the ether extract fraction, varies depending upon the plant part. True fat and oil (triglycerides) occur only in storage organs such as seeds; the leaves and stems are virtually free of triglycerides. The fatty acid fractions of leaves and stems are contained in galactolipids of lower energy content than triglycerides. Leaves and stems also contain waxes, chlorophyll, essential oils (esters and terpenes), pigments, saponins, flavonoids, isoflavonoids, and alkaloids, most of which have no nutritive value or inhibit the utilization of feed. Utilizable fatty acids constitute no more than 50 percent of forage lipids, but are the main component of seeds and grain by-products.

LINOLEIC ACID

Values for linoleic acid, an essential fatty acid, are shown in Table 5, where the information was available. The major sources of linoleic acid in feedstuffs are the vegetable oils. Corn oil and cottonseed oil are approximately 50 percent linoleic acid; safflower oil is 75 percent linoleic acid. Yellow corn is the major source of linoleic acid in many feed formulas.

MINERALS

Values for the important mineral elements are shown in Table 2. Several other minerals known or thought to be required are not listed because of paucity of compositional data. The values shown are the total percentage or weight of the mineral present. The availability (digestibility) of minerals in feedstuffs varies considerably and can be an important factor in the value of a feed as a source of a particular mineral for animals.

The composition of mineral supplements is shown in Table 6.

VITAMINS

Values for some important vitamins are shown in Table 3. Xanthophyll, which is useful in poultry diets to provide yellow coloration in egg yolks and yellow skin coloration, is listed in this table although it is not a vitamin. Carotene (provitamin A) values are provided but vitamin A values are not, because species convert carotene to vitamin A at different rates (see Table 12). Vitamin A standards are as follows:

The international standard for vitamin A activity as related to vitamin A and beta-carotene are as follows:

$$
\begin{aligned}
1 \text{ IU vitamin A} &= 1 \text{ USP unit} \\
&= \text{vitamin A activity of 0.300 microgram} \\
&\quad \text{crystalline vitamin A alcohol} \\
&= \text{vitamin A activity of 0.344 microgram} \\
&\quad \text{vitamin A acetate} \\
&= \text{vitamin A activity of 0.550 microgram} \\
&\quad \text{vitamin A palmitate} \\
1 \text{ IU vitamin A} &= 0.6 \text{ microgram beta-carotene} \\
1 \text{ mg beta-carotene} &= 1,667 \text{ IU vitamin A.}
\end{aligned}
$$

International standards for vitamin A are based on the utilization of vitamin A and beta-carotene by the rat. Because the various species of animals do not convert carotene to vitamin A in the same ratio as rats, it is suggested that conversion rates in Table 12 be used.

A detailed discussion of the variations in vitamin activity and nomenclature is beyond the scope of this publication. Compounds with different levels of vitamin D, E, and K activity are known to occur in nature. The complexity of vitamin nomenclature precludes incorporating variations in a single vitamin table. For instance, folacin and folic acid are frequently used interchangeably, but folacin is the correct term for describing the activity of this vitamin in feedstuffs. Likewise, vitamin B_6 refers to a complete class of three compounds; whereas, pyridoxine refers specifically to the primary alcohol form.

FEED
COMPOSITION
TABLES

TABLE 1   Composition of Important Feeds: Energy Values, Proximate Analyses, Plant Cell Wall Constituents, and Acid Detergent Fiber, Data Expressed As-Fed and Dry (100% Dry Matter)

Entry Number	Feed Name Description	International Feed Number	Dry Matter (%)	Ruminants TDN (%)	DE (Mcal/kg)	ME (Mcal/kg)	NE_m (Mcal/kg)	NE_g (Mcal/kg)	Dairy Cattle NE_l (Mcal/kg)	Chickens ME_n (kcal/kg)	TME (kcal/kg)	NE_p (kcal/kg)
	ALFALFA *Medicago sativa*											
001	fresh, late vegetative	2-00-181	21.0	13.0	0.59	0.50	0.30	0.16	0.30	—	—	—
002			100.0	63.0	2.78	2.36	1.39	0.75	1.42	—	—	—
003	fresh, early bloom	2-00-184	23.0	14.0	0.61	0.51	0.30	0.15	0.31	—	—	—
004			100.0	60.0	2.65	2.22	1.31	0.65	1.35	—	—	—
005	fresh, midbloom	2-00-185	24.0	14.0	0.62	0.52	0.31	0.14	0.32	—	—	—
006			100.0	58.0	2.56	2.13	1.26	0.58	1.30	—	—	—
007	fresh, full bloom	2-00-188	25.0	14.0	0.61	0.50	0.30	0.12	0.31	—	—	—
008			100.0	55.0	2.43	2.00	1.19	0.47	1.23	—	—	—
009	hay, sun-cured, late bloom	1-20-681	90.0	47.0	2.06	1.68	1.01	0.32	1.04	—	—	—
010			100.0	52.0	2.29	1.87	1.12	0.36	1.15	—	—	—
011	hay, sun-cured, mature	1-00-071	91.0	46.0	2.01	1.62	0.98	0.25	1.01	—	—	—
012			100.0	50.0	2.21	1.78	1.07	0.28	1.11	—	—	—
013	leaves, sun-cured	1-00-146	89.0	64.0	2.84	2.46	1.46	0.92	1.47	—	—	—
014			100.0	72.0	3.17	2.76	1.64	1.03	1.64	—	—	—
015	meal dehy, 15% protein	1-00-022	90.0	54.0	2.35	1.97	1.16	0.56	1.20	1,535.0	1,094.0	525.0
016			100.0	59.0	2.60	2.18	1.28	0.62	1.33	1,698.0	1,209.0	581.0
017	meal dehy, 17% protein	1-00-023	92.0	55.0	2.47	2.08	1.22	0.63	1.26	1,504.0	1,393.0	770.0
018			100.0	61.0	2.69	2.27	1.33	0.69	1.38	1,640.0	1,519.0	840.0
019	meal dehy, 20% protein	1-00-024	92.0	57.0	2.51	2.12	1.25	0.66	1.28	1,625.0	1,429.0	1,020.0
020			100.0	62.0	2.73	2.31	1.36	0.72	1.40	1,774.0	1,560.0	1,113.0
021	meal dehy, 22% protein	1-07-851	93.0	62.0	2.74	2.35	1.39	0.82	1.41	1,692.0	1,661.0	1,155.0
022			100.0	67.0	2.95	2.53	1.50	0.88	1.52	1,823.0	1,790.0	1,245.0
023	wilted silage, early bloom	3-00-216	35.0	21.0	0.92	0.77	0.45	0.23	0.47	—	—	—
024			100.0	60.0	2.65	2.22	1.31	0.65	1.35	—	—	—
025	wilted silage, midbloom	3-00-217	38.0	22.0	0.97	0.81	0.48	0.22	0.50	—	—	—
026			100.0	58.0	2.56	2.13	1.26	0.58	1.30	—	—	—
027	wilted silage, full bloom	3-00-218	45.0	25.0	1.09	0.90	0.53	0.21	0.55	—	—	—
028			100.0	55.0	2.43	2.00	1.19	0.47	1.23	—	—	—
	NORTH											
029	hay, sun-cured, early	1-00-050	90.0	59.0	2.62	2.24	1.32	0.76	1.35	—	—	—
030	vegetative		100.0	66.0	2.91	2.49	1.47	0.85	1.50	—	—	—
031	hay, sun-cured, late	1-00-054	90.0	57.0	2.49	2.11	1.24	0.67	1.28	—	—	—
032	vegetative		100.0	63.0	2.78	2.36	1.39	0.75	1.42	—	—	—
033	hay, sun-cured, early bloom	1-00-059	90.0	54.0	2.38	2.00	1.18	0.59	1.22	—	—	—
034			100.0	60.0	2.65	2.22	1.31	0.65	1.35	—	—	—
035	hay, sun-cured, midbloom	1-00-063	90.0	52.0	2.30	1.92	1.13	0.52	1.17	—	—	—
036			100.0	58.0	2.56	2.13	1.26	0.58	1.30	—	—	—
037	hay, sun-cured, full bloom	1-00-068	90.0	50.0	2.18	1.80	1.07	0.43	1.11	—	—	—
038			100.0	55.0	2.43	2.00	1.19	0.47	1.23	—	—	—
	SOUTH											
039	hay, sun-cured, early vege-	1-00-050	90.0	59.0	2.58	2.20	1.30	0.74	1.33	—	—	—
040	tative		100.0	65.0	2.87	2.45	1.44	0.82	1.47	—	—	—
041	hay, sun-cured, late vege-	1-00-054	90.0	57.0	2.49	2.11	1.24	0.67	1.28	—	—	—
042	tative		100.0	63.0	2.78	2.36	1.39	0.75	1.42	—	—	—
043	hay, sun-cured, early bloom	1-00-059	90.0	53.0	2.34	1.96	1.16	0.56	1.19	—	—	—
044			100.0	59.0	2.60	2.18	1.28	0.62	1.33	—	—	—
045	hay, sun-cured, midbloom	1-00-063	90.0	51.0	2.26	1.88	1.11	0.49	1.15	—	—	—
046			100.0	57.0	2.51	2.09	1.23	0.55	1.28	—	—	—
047	hay, sun-cured, full bloom	1-00-068	90.0	49.0	2.14	1.76	1.05	0.39	1.08	—	—	—
048			100.0	54.0	2.38	1.96	1.16	0.43	1.20	—	—	—

| | Horses | | | Swine | | | | Plant Cell Wall Constituents | | | | | | | |
Entry Num-ber	TDN (%)	DE (Mcal/ kg)	ME (Mcal/ kg)	TDN (%)	DE (kcal/ kg)	ME (kcal/ kg)	Crude Pro-tein (%)	Cell Walls (%)	Cell-ulose (%)	Hemi-cell-ulose (%)	Lig-nin (%)	Acid Deter-gent Fiber (%)	Crude Fiber (%)	Ether Ex-tract (%)	Ash (%)
001	—	—	—	12.0	548.0	502.0	4.3	8.0	5.0	1.0	1.0	6.0	4.9	0.6	2.1
002	—	—	—	58.0	2,566.0	2,351.0	20.0	38.0	22.0	7.0	7.0	29.0	23.0	2.7	9.8
003	—	—	—	—	—	—	4.4	9.0	5.0	2.0	2.0	7.0	5.8	0.7	2.2
004	—	—	—	—	—	—	19.0	40.0	23.0	8.0	7.0	31.0	25.0	3.1	9.5
005	—	—	—	14.0	631.0	581.0	4.5	11.0	6.0	2.0	2.0	9.0	6.8	0.6	2.1
006	—	—	—	59.0	2,583.0	2,379.0	18.3	46.0	26.0	10.0	9.0	35.0	28.0	2.6	8.7
007	—	—	—	—	—	—	3.5	13.0	7.0	3.0	2.0	9.0	7.7	0.7	2.1
008	—	—	—	—	—	—	14.0	52.0	27.0	13.0	10.0	37.0	31.0	2.8	8.5
009	—	—	—	—	—	—	12.6	47.0	23.0	11.0	11.0	35.0	28.8	1.6	7.0
010	—	—	—	—	—	—	14.0	52.0	26.0	12.0	12.0	39.0	32.0	1.8	7.8
011	42.0	1.68	1.38	—	—	—	11.7	53.0	26.0	12.0	13.0	40.0	34.4	1.2	6.9
012	46.0	1.84	1.51	—	—	—	12.9	58.0	29.0	13.0	14.0	44.0	37.7	1.3	7.5
013	52.0	2.04	1.68	—	—	—	20.6	30.0	14.0	5.0	4.0	21.0	15.8	2.7	9.6
014	58.0	2.29	1.88	—	—	—	23.1	34.0	16.0	6.0	5.0	24.0	17.7	3.0	10.7
015	46.0	1.83	1.50	31.0	1,372.0	1,293.0	15.6	46.0	26.0	—	11.0	37.0	26.6	2.2	9.1
016	51.0	2.03	1.66	34.0	1,517.0	1,430.0	17.3	51.0	29.0	—	12.0	41.0	29.4	2.5	10.0
017	45.0	1.79	1.47	44.0	1,418.0	1,196.0	17.3	41.0	22.0	—	10.0	32.0	24.0	2.7	9.7
018	49.0	1.95	1.60	48.0	1,546.0	1,304.0	18.9	45.0	24.0	—	11.0	35.0	26.2	3.0	10.6
019	38.0	1.55	1.27	48.0	2,080.0	1,923.0	20.2	38.0	20.0	—	7.0	28.0	20.6	3.3	10.4
020	42.0	1.69	1.39	52.0	2,270.0	2,099.0	22.0	42.0	22.0	—	8.0	31.0	22.5	3.7	11.3
021	27.0	1.14	0.94	49.0	2,186.0	1,855.0	22.2	36.0	19.0	—	7.0	26.0	18.3	4.1	10.2
022	29.0	1.23	1.01	53.0	2,355.0	1,999.0	23.9	39.0	20.0	—	8.0	28.0	19.8	4.4	11.0
023	—	—	—	—	—	—	5.9	15.0	8.0	3.0	3.0	11.0	9.7	1.1	2.8
024	—	—	—	—	—	—	17.0	43.0	23.0	9.0	10.0	33.0	28.0	3.2	8.2
025	—	—	—	—	—	—	5.9	18.0	9.0	4.0	4.0	13.0	11.4	1.2	3.0
026	—	—	—	—	—	—	15.5	47.0	24.0	10.0	11.0	35.0	30.0	3.1	7.9
027	—	—	—	—	—	—	6.3	23.0	11.0	5.0	5.0	17.0	14.9	1.2	3.5
028	—	—	—	—	—	—	14.0	51.0	25.0	12.0	12.0	38.0	33.2	2.7	7.7
029	—	—	—	—	—	—	20.7	34.0	20.0	6.0	5.0	25.0	18.4	3.6	9.2
030	—	—	—	—	—	—	23.0	38.0	22.0	7.0	5.0	28.0	20.5	4.0	10.2
031	49.0	1.93	1.58	—	—	—	17.9	36.0	21.0	7.0	6.0	26.0	19.7	3.4	8.3
032	54.0	2.15	1.76	—	—	—	20.0	40.0	23.0	8.0	7.0	29.0	22.0	3.8	9.2
033	46.0	1.83	1.50	—	—	—	16.2	38.0	22.0	8.0	7.0	28.0	20.7	2.7	8.6
034	51.0	2.04	1.67	—	—	—	18.0	42.0	24.0	9.0	8.0	31.0	23.0	3.0	9.6
035	42.0	1.70	1.39	—	—	—	15.3	41.0	23.0	9.0	8.0	32.0	23.4	2.3	8.2
036	47.0	1.89	1.55	—	—	—	17.0	46.0	26.0	10.0	9.0	35.0	26.0	2.6	9.1
037	45.0	1.78	1.46	—	—	—	13.5	45.0	25.0	10.0	9.0	33.0	26.1	1.8	8.0
038	50.0	1.98	1.62	—	—	—	15.0	50.0	28.0	11.0	10.0	37.0	29.0	2.0	8.9
039	—	—	—	—	—	—	25.2	31.0	22.0	5.0	5.0	23.0	17.1	3.6	9.2
040	—	—	—	—	—	—	28.0	34.0	25.0	6.0	6.0	26.0	19.0	4.0	10.2
041	49.0	1.93	1.58	—	—	—	22.4	37.0	27.0	6.0	7.0	24.0	17.9	3.4	8.3
042	54.0	2.15	1.76	—	—	—	25.0	42.0	31.0	7.0	8.0	27.0	20.0	3.8	9.2
043	46.0	1.83	1.50	—	—	—	20.7	36.0	18.0	7.0	8.0	27.0	20.7	3.2	8.0
044	51.0	2.04	1.67	—	—	—	23.0	40.0	20.0	8.0	9.0	30.0	23.0	3.6	8.9
045	42.0	1.70	1.39	—	—	—	17.1	40.0	19.0	9.0	9.0	29.0	22.5	2.9	7.6
046	47.0	1.89	1.55	—	—	—	19.0	44.0	21.0	10.0	10.0	32.0	25.0	3.2	8.5
047	45.0	1.78	1.46	—	—	—	15.3	43.0	23.0	10.0	10.0	32.0	24.3	2.0	7.3
048	50.0	1.98	1.62	—	—	—	17.0	48.0	26.0	11.0	11.0	36.0	27.0	2.2	8.1

TABLE 1 Composition of Important Feeds: Energy Values, Proximate Analyses, Plant Cell Wall Constituents, and Acid Detergent Fiber—*Continued*

Entry Number	Feed Name Description	International Feed Number	Dry Matter (%)	Ruminants TDN (%)	DE (Mcal/kg)	ME (Mcal/kg)	NE_m (Mcal/kg)	NE_g (Mcal/kg)	Dairy Cattle NE_l (Mcal/kg)	Chickens ME_n (kcal/kg)	TME (kcal/kg)	NE_p (kcal/kg)
	ALMOND *Prunus amygdalus*											
049	hulls	4-00-359	90.0	54.0	2.38	2.00	1.18	0.59	1.21	—	—	—
050			100.0	60.0	2.65	2.22	1.31	0.65	1.35	—	—	—
	ANIMAL											
051	by-product, meal rendered	5-08-786	93.0	68.0	2.99	2.61	1.55	0.98	1.55	2,714.0	—	—
052			100.0	73.0	3.22	2.80	1.67	1.06	1.67	2,918.0	—	—
	APPLES *Malus* spp											
053	pomace, oat hulls added,	4-28-096	89.0	60.0	2.67	2.30	1.36	0.81	1.38	—	1,746.0	—
054	dehy		100.0	68.0	3.00	2.58	1.52	0.91	1.55	—	1,960.0	—
	BAHIAGRASS *Paspalum notatum*											
055	fresh	2-00-464	30.0	16.0	0.70	0.58	0.34	0.13	0.36	—	—	—
056			100.0	54.0	2.38	1.96	1.16	0.43	1.20	—	—	—
057	hay, sun-cured	1-00-462	91.0	46.0	2.05	1.66	1.00	0.29	1.03	—	—	—
058			100.0	51.0	2.25	1.82	1.10	0.32	1.13	—	—	—
059	hay, sun-cured, early vege-	1-06-137	91.0	44.0	1.93	1.54	0.94	0.17	0.96	—	—	—
060	tative		100.0	48.0	2.12	1.69	1.03	0.19	1.06	—	—	—
061	hay, sun-cured, late vege-	1-20-787	91.0	40.0	1.77	1.37	0.87	0.01	0.87	—	—	—
062	tative		100.0	44.0	1.94	1.51	0.96	0.01	0.96	—	—	—
063	hay, sun-cured, early bloom	1-06-138	91.0	36.0	1.61	1.21	0.81	—	0.78	—	—	—
064			100.0	40.0	1.76	1.33	0.89	—	0.86	—	—	—
	BAKERY											
065	waste, dehy (Dried bakery	4-00-466	92.0	82.0	3.61	3.23	1.98	1.35	1.89	3,862.0	—	2,879.0
066	product)		100.0	89.0	3.92	3.51	2.15	1.47	2.06	4,203.0	—	3,133.0
	BARLEY *Hordeum vulgare*											
067	grain	4-00-549	88.0	74.0	3.27	2.90	1.76	1.19	1.71	2,508.0	3,011.0	1,803.0
068			100.0	84.0	3.70	3.29	2.00	1.35	1.94	2,843.0	3,413.0	2,044.0
069	grain, light 46.3 kg/hl	4-00-566	89.0	69.0	3.02	2.65	1.59	1.04	1.57	—	—	—
070	(Less than 36 lb/bushel)		100.0	77.0	3.40	2.98	1.79	1.17	1.77	—	—	—
071	grain, Pacific Coast	4-07-939	89.0	77.0	3.38	3.01	1.84	1.25	1.77	2,598.0	3,083.0	1,809.0
072			100.0	86.0	3.79	3.38	2.06	1.40	1.99	2,914.0	3,457.0	2,029.0
073	grain screenings	4-00-542	89.0	71.0	3.14	2.77	1.67	1.11	1.64	1,797.0	1,796.0	—
074			100.0	80.0	3.53	3.11	1.88	1.24	1.84	2,021.0	2,020.0	—
075	hay, sun-cured	1-00-495	87.0	49.0	2.16	1.79	1.06	0.45	1.09	—	—	—
076			100.0	56.0	2.47	2.04	1.21	0.51	1.25	—	—	—
077	malt sprouts, dehy	5-00-545	94.0	66.0	2.93	2.54	1.51	0.94	1.52	1,463.0	—	1,205.0
078			100.0	71.0	3.13	2.71	1.61	1.00	1.62	1,561.0	—	1,286.0
079	straw	1-00-498	91.0	45.0	1.97	1.58	0.96	0.21	0.99	—	—	—
080			100.0	49.0	2.16	1.73	1.05	0.23	1.08	—	—	—
	BEAN, NAVY *Phaseolus vulgaris*											
081	seeds	5-00-623	89.0	75.0	3.31	2.95	1.79	1.21	1.73	2,320.0	—	986.0
082			100.0	84.0	3.70	3.29	2.00	1.35	1.94	2,593.0	—	1,102.0
	BEET, MANGELS *Beta vulgaris macrorhiza*											
083	roots, fresh	4-00-637	11.0	9.0	0.39	0.34	0.21	0.14	0.20	—	—	—
084			100.0	80.0	3.53	3.11	1.88	1.24	1.84	—	—	—

	Horses			Swine				Plant Cell Wall Constituents							
Entry Num- ber	TDN (%)	DE (Mcal/ kg)	ME (Mcal/ kg)	TDN (%)	DE (kcal/ kg)	ME (kcal/ kg)	Crude Pro- tein (%)	Cell Walls (%)	Cell- ulose (%)	Hemi- cell- ulose (%)	Lig- nin (%)	Acid Deter- gent Fiber (%)	Crude Fiber (%)	Ether Ex- tract (%)	Ash (%)
049	—	—	—	68.0	2,986.0	2,838.0	1.9	29.0	—	—	8.0	25.0	13.5	2.7	5.8
050	—	—	—	75.0	3,323.0	3,158.0	2.1	32.0	—	—	9.0	28.0	15.0	3.0	6.5
051	—	—	—	—	—	—	60.2	—	—	—	—	—	2.2	9.0	21.9
052	—	—	—	—	—	—	64.7	—	—	—	—	—	2.4	9.7	23.6
053	—	—	—	63.0	2,778.0	2,638.0	4.6	—	—	—	12.0	40.0	17.8	4.7	3.1
054	—	—	—	71.0	3,118.0	2,961.0	5.1	—	—	—	14.0	45.0	20.0	5.2	3.5
055	—	—	—	—	—	—	2.6	20.0	—	—	2.0	11.0	9.0	0.5	3.3
056	—	—	—	—	—	—	8.9	68.0	—	—	7.0	38.0	30.4	1.6	11.1
057	40.0	1.61	1.32	—	—	—	7.4	66.0	29.0	27.0	7.0	37.0	29.2	1.9	5.9
058	44.0	1.77	1.45	—	—	—	8.2	72.0	32.0	30.0	8.0	41.0	32.0	2.1	6.4
059	—	—	—	—	—	—	10.9	64.0	25.0	—	4.0	30.0	26.4	1.7	9.1
060	—	—	—	—	—	—	12.0	70.0	28.0	—	4.0	33.0	29.0	1.9	10.0
061	—	—	—	—	—	—	8.6	66.0	28.0	—	5.0	35.0	30.0	1.5	8.7
062	—	—	—	—	—	—	9.5	73.0	31.0	—	6.0	38.0	33.0	1.7	9.6
063	—	—	—	—	—	—	6.4	69.0	31.0	—	6.0	38.0	30.9	1.4	8.5
064	—	—	—	—	—	—	7.0	76.0	34.0	—	7.0	42.0	34.0	1.5	9.3
065	—	—	—	90.0	3,983.0	3,738.0	9.8	—	—	—	—	—	1.2	11.7	4.0
066	—	—	—	98.0	4,335.0	4,068.0	10.7	—	—	—	—	—	1.3	12.7	4.4
067	72.0	—	—	70.0	3,108.0	2,910.0	11.9	17.0	4.0	—	2.0	6.0	5.0	1.9	2.3
068	82.0	—	—	79.0	3,523.0	3,299.0	13.5	19.0	5.0	—	2.0	7.0	5.7	2.1	2.6
069	—	—	—	71.0	3,116.0	2,903.0	12.4	—	—	—	—	—	7.7	1.9	3.3
070	—	—	—	79.0	3,498.0	3,259.0	14.0	—	—	—	—	—	8.6	2.2	3.7
071	—	—	—	71.0	3,127.0	2,937.0	9.6	19.0	—	—	—	8.0	6.3	1.8	2.7
072	—	—	—	80.0	3,507.0	3,293.0	10.8	21.0	—	—	—	9.0	7.1	2.0	3.1
073	—	—	—	70.0	3,092.0	2,330.0	11.7	—	—	—	—	—	8.6	2.3	3.1
074	—	—	—	79.0	3,478.0	2,621.0	13.1	—	—	—	—	—	9.6	2.6	3.4
075	39.0	1.56	1.28	—	—	—	7.6	—	—	—	—	—	24.1	1.9	6.6
076	44.0	1.79	1.47	—	—	—	8.7	—	—	—	—	—	27.5	2.1	7.6
077	—	—	—	36.0	2,438.0	2,204.0	26.3	44.0	13.0	25.0	3.0	17.0	15.0	1.3	6.6
078	—	—	—	38.0	2,602.0	2,352.0	28.1	47.0	14.0	27.0	3.0	18.0	16.0	1.4	7.0
079	34.0	1.40	1.15	—	—	—	4.0	73.0	34.0	40.0	10.0	54.0	38.3	1.7	6.5
080	37.0	1.54	1.26	—	—	—	4.3	80.0	37.0	44.0	11.0	59.0	42.0	1.9	7.1
081	—	—	—	—	3,714.0	3,376.0	22.6	—	—	—	—	—	4.5	1.3	4.7
082	—	—	—	—	4,150.0	3,772.0	25.3	—	—	—	—	—	5.0	1.5	5.2
083	—	—	—	9.0	399.0	374.0	1.3	—	—	—	—	—	0.8	0.1	1.1
084	—	—	—	83.0	3,642.0	3,410.0	11.8	—	—	—	—	—	7.4	0.7	9.6

TABLE 1   Composition of Important Feeds: Energy Values, Proximate Analyses, Plant Cell Wall Constituents, and Acid Detergent Fiber—*Continued*

Entry Number	Feed Name Description	International Feed Number	Dry Matter (%)	Ruminants TDN (%)	DE (Mcal/kg)	ME (Mcal/kg)	NE$_m$ (Mcal/kg)	NE$_g$ (Mcal/kg)	Dairy Cattle NE$_l$ (Mcal/kg)	Chickens ME$_n$ (kcal/kg)	TME (kcal/kg)	NE$_p$ (kcal/kg)
	BEET, SUGAR *Beta vulgaris altissima*											
085	aerial part with crowns,	3-00-660	22.0	11.0	0.50	0.41	0.25	0.07	0.25	—	—	—
086	silage		100.0	51.0	2.25	1.82	1.10	0.32	1.13	—	—	—
	molasses—see Molasses and syrup											
087	pulp, dehy	4-00-669	91.0	67.0	2.96	2.58	1.54	0.98	1.54	646.0	—	438.0
088			100.0	74.0	3.26	2.85	1.70	1.08	1.69	713.0	—	483.0
089	pulp, wet	4-00-671	11.0	9.0	0.38	0.33	0.20	0.13	0.20	—	—	—
090			100.0	78.0	3.44	3.02	1.82	1.19	1.79	—	—	—
091	pulp with molasses, dehy	4-00-672	92.0	70.0	3.07	2.69	1.61	1.04	1.60	659.0	2,141.0	440.0
092			100.0	76.0	3.35	2.93	1.76	1.14	1.74	719.0	2,334.0	479.0
093	pulp with steffens filtrate,	4-00-675	92.0	61.0	2.66	2.28	1.34	0.78	1.37	—	—	—
094	dehy (Dried beet product)		100.0	66.0	2.91	2.49	1.47	0.85	1.50	—	—	—
	BENTGRASS, CREEPING *Agrostis palustris*											
095	hay, sun-cured, postripe	1-00-688	92.0	48.0	2.11	1.72	1.03	0.33	1.06	—	—	—
096			100.0	52.0	2.29	1.87	1.12	0.36	1.15	—	—	—
	BERMUDAGRASS *Cynodon dactylon*											
097	fresh	2-00-712	34.0	20.0	0.89	0.75	0.44	0.22	0.45	—	—	—
098			100.0	60.0	2.65	2.22	1.31	0.65	1.35	—	—	—
099	hay, sun-cured	1-00-703	91.0	42.0	1.85	1.46	0.91	0.10	0.92	—	—	—
100			100.0	46.0	2.03	1.60	0.99	0.10	1.01	—	—	—
	BERMUDAGRASS, COASTAL *Cynodon dactylon*											
101	fresh	2-00-719	29.0	19.0	0.82	0.70	0.41	0.23	0.42	—	—	—
102			100.0	64.0	2.82	2.40	1.41	0.78	1.45	—	—	—
103	hay, sun-cured	1-00-716	90.0	49.0	2.15	1.77	1.05	0.39	1.09	—	—	—
104			100.0	54.0	2.38	1.96	1.16	0.43	1.20	—	—	—
105	hay, sun-cured, early vege-	1-00-713	94.0	58.0	2.53	2.13	1.25	0.64	1.29	—	—	—
106	tative		100.0	61.0	2.69	2.27	1.33	0.69	1.38	—	—	—
107	hay, sun-cured, late vege-	1-20-900	91.0	50.0	2.17	1.78	1.06	0.40	1.10	—	—	—
108	tative		100.0	54.0	2.38	1.96	1.16	0.43	1.20	—	—	—
109	hay, sun-cured, 15 to 28	1-09-207	92.0	51.0	2.23	1.84	1.09	0.43	1.13	—	—	—
110	days' growth		100.0	55.0	2.43	2.00	1.19	0.47	1.23	—	—	—
111	hay, sun-cured, 29 to 42	1-09-209	93.0	47.0	2.05	1.65	1.00	0.26	1.03	—	—	—
112	days' growth		100.0	50.0	2.21	1.78	1.07	0.28	1.11	—	—	—
113	hay, sun-cured, 43 to 56	1-09-210	93.0	40.0	1.76	1.36	0.87	—	0.87	—	—	—
114	days' growth		100.0	43.0	1.90	1.47	0.94	—	0.93	—	—	—
	BERMUDAGRASS, COAST-CROSS *Cynodon dactylon*											
115	hay, sun-cured	1-28-254	90.0	48.0	2.10	1.72	1.03	0.36	1.06	—	—	—
116			100.0	53.0	2.34	1.91	1.14	0.40	1.18	—	—	—
	BERMUDAGRASS, MID-LAND *Cynodon dactylon*											
117	hay, sun-cured, 15 to 28	1-06-139	92.0	49.0	2.15	1.76	1.05	0.36	1.08	—	—	—
118	days' growth (South)		100.0	53.0	2.34	1.91	1.14	0.40	1.18	—	—	—
119	hay, sun-cured, 29 to 42	1-06-140	92.0	44.0	1.95	1.55	0.95	0.18	0.97	—	—	—
120	days' growth (South)		100.0	48.0	2.12	1.69	1.03	0.19	1.06	—	—	—

	Horses			Swine				Plant Cell Wall Constituents							
Entry Number	TDN (%)	DE (Mcal/ kg)	ME (Mcal/ kg)	TDN (%)	DE (kcal/ kg)	ME (kcal/ kg)	Crude Protein (%)	Cell Walls (%)	Cell-ulose (%)	Hemi-cell-ulose (%)	Lig-nin (%)	Acid Deter-gent Fiber (%)	Crude Fiber (%)	Ether Ex-tract (%)	Ash (%)
085	—	—	—	—	—	—	3.0	—	—	—	—	—	3.1	0.6	7.3
086	—	—	—	—	—	—	13.4	—	—	—	—	—	13.7	2.8	32.5
087	—	—	—	67.0	2,932.0	2,693.0	8.8	49.0	—	—	2.0	30.0	18.0	0.5	4.9
088	—	—	—	73.0	3,235.0	2,971.0	9.7	54.0	—	—	2.0	33.0	19.8	0.6	5.4
089	—	—	—	—	—	—	1.2	—	—	—	—	—	3.1	0.2	0.5
090	—	—	—	—	—	—	11.2	—	—	—	—	—	28.1	2.1	4.7
091	—	—	—	70.0	3,066.0	2.880.0	9.3	40.0	—	—	3.0	23.0	15.1	0.6	5.6
092	—	—	—	76.0	3,342.0	3,139.0	10.1	44.0	—	—	3.0	25.0	16.5	0.6	6.1
093	—	—	—	58.0	2,570.0	2,404.0	16.5	38.0	—	—	2.0	19.0	15.2	0.3	5.6
094	—	—	—	64.0	2,808.0	2,626.0	18.0	42.0	—	—	2.0	21.0	16.6	0.4	6.1
095	31.0	1.31	1.07	—	—	—	4.0	—	—	—	—	—	28.3	1.3	5.3
096	34.0	1.42	1.16	—	—	—	4.3	—	—	—	—	—	30.8	1.4	5.8
097	—	—	—	23.0	1,019.0	954.0	4.1	—	—	—	—	—	8.9	0.8	3.4
098	—	—	—	69.0	3,030.0	2,835.0	12.0	—	—	—	—	—	26.4	2.2	10.2
099	41.0	1.67	1.37	—	—	—	8.9	—	—	—	—	—	27.8	1.8	8.4
100	45.0	1.82	1.50	—	—	—	9.8	—	—	—	—	—	30.4	2.0	9.2
101	—	—	—	—	—	—	4.4	—	—	—	—	—	8.3	1.1	1.8
102	—	—	—	—	—	—	15.0	—	—	—	—	—	28.4	3.8	6.3
103	43.0	1.73	1.42	—	—	—	5.4	70.0	—	—	5.0	34.0	27.7	2.1	5.9
104	48.0	1.92	1.57	—	—	—	6.0	78.0	—	—	6.0	38.0	30.7	2.3	6.6
105	46.0	1.84	1.51	—	—	—	15.0	62.0	—	—	4.0	29.0	25.2	2.4	5.7
106	49.0	1.96	1.61	—	—	—	16.0	66.0	—	—	4.0	30.0	26.8	2.5	6.1
107	46.0	1.84	1.51	—	—	—	15.0	—	—	—	—	—	24.8	1.6	7.0
108	51.0	2.02	1.66	—	—	—	16.5	—	—	—	—	—	27.3	1.8	7.7
109	—	—	—	—	—	—	14.7	68.0	26.0	—	4.0	30.0	24.8	2.6	10.1
110	—	—	—	—	—	—	16.0	74.0	28.0	—	4.0	33.0	27.0	2.8	11.0
111	—	—	—	—	—	—	11.2	71.0	28.0	—	5.0	35.0	30.7	2.0	9.3
112	—	—	—	—	—	—	12.0	76.0	30.0	—	6.0	38.0	33.0	2.1	10.0
113	—	—	—	—	—	—	7.4	73.0	31.0	—	7.0	40.0	33.5	1.3	8.4
114	—	—	—	—	—	—	8.0	78.0	33.0	—	7.0	43.0	36.0	1.4	9.0
115	—	—	—	—	—	—	7.7	—	—	—	—	—	31.4	—	—
116	—	—	—	—	—	—	8.5	—	—	—	—	—	34.9	—	—
117	—	—	—	—	—	—	14.0	65.0	24.0	—	4.0	30.0	25.8	2.5	11.8
118	—	—	—	—	—	—	15.2	71.0	26.0	—	5.0	33.0	28.0	2.7	12.8
119	—	—	—	—	—	—	9.3	71.0	26.0	—	5.0	35.0	27.6	1.8	11.4
120	—	—	—	—	—	—	10.1	77.0	28.0	—	5.0	38.0	30.0	2.0	12.4

TABLE 1 Composition of Important Feeds: Energy Values, Proximate Analyses, Plant Cell Wall Constituents, and Acid Detergent Fiber—*Continued*

Entry Number	Feed Name Description	International Feed Number	Dry Matter (%)	Ruminants TDN (%)	DE (Mcal/kg)	ME (Mcal/kg)	NE$_m$ (Mcal/kg)	NE$_g$ (Mcal/kg)	Dairy Cattle NE$_l$ (Mcal/kg)	Chickens ME$_n$ (kcal/kg)	TME (kcal/kg)	NE$_p$ (kcal/kg)
	BIRDSFOOT TREFOIL—SEE TREFOIL, BIRDSFOOT											
	BLOOD											
121	meal	5-00-380	92.0	61.0	2.66	2.28	1.34	0.78	1.37	2,833.0	2,361.0	2,179.0
122			100.0	66.0	2.91	2.49	1.47	0.85	1.50	3,096.0	2,580.0	2,381.0
123	meal flash dehy	5-26-006	92.0	60.0	2.64	2.25	1.33	0.75	1.36	2,821.0	3,200.0	2,241.0
124			100.0	65.0	2.87	2.45	1.44	0.82	1.47	3,065.0	3,476.0	2,435.0
125	meal spray dehy (Blood	5-00-381	93.0	85.0	3.73	3.35	2.06	1.41	1.96	2,784.0	—	2,251.0
126	flour)		100.0	91.0	4.01	3.60	2.22	1.52	2.11	2,994.0	—	2,420.0
	BLUEGRASS, CANADA *Poa compressa*											
127	fresh, early vegetative	2-00-763	26.0	18.0	0.81	0.70	0.42	0.26	0.42	—	—	—
128			100.0	71.0	3.13	2.71	1.61	1.00	1.62	—	—	—
129	hay, sun-cured, late vege-	1-20-889	97.0	69.0	3.04	2.63	1.56	0.97	1.57	—	—	—
130	tative		100.0	71.0	3.13	2.71	1.61	1.00	1.62	—	—	—
	BLUEGRASS, KENTUCKY *Poa pratensis*											
131	fresh	2-00-786	35.0	23.0	1.00	0.85	0.50	0.28	0.51	—	—	—
132			100.0	64.0	2.82	2.40	1.41	0.78	1.45	—	—	—
133	fresh, early vegetative	2-00-777	31.0	22.0	0.98	0.85	0.51	0.32	0.51	—	—	—
134			100.0	72.0	3.17	2.76	1.64	1.03	1.64	—	—	—
135	fresh, early bloom	2-00-779	35.0	24.0	1.07	0.92	0.54	0.33	0.55	—	—	—
136			100.0	69.0	3.04	2.62	1.55	0.94	1.57	—	—	—
137	fresh, milk stage	2-00-782	42.0	27.0	1.17	0.99	0.58	0.32	0.60	—	—	—
138			100.0	63.0	2.78	2.36	1.39	0.75	1.42	—	—	—
139	fresh, mature	2-00-784	42.0	23.0	1.03	0.85	0.50	0.21	0.52	—	—	—
140			100.0	56.0	2.47	2.04	1.21	0.51	1.25	—	—	—
141	hay, sun-cured	1-00-776	89.0	54.0	2.40	2.02	1.19	0.61	1.22	—	—	—
142			100.0	61.0	2.69	2.27	1.33	0.69	1.38	—	—	—
143	hay, sun-cured, full bloom	1-00-772	92.0	52.0	2.32	1.92	1.14	0.50	1.18	—	—	—
144			100.0	57.0	2.51	2.09	1.23	0.55	1.28	—	—	—
145	silage, early bloom	3-00-788	41.0	25.0	1.10	0.92	0.54	0.27	0.56	—	—	—
146			100.0	60.0	2.65	2.22	1.31	0.65	1.35	—	—	—
	BLUESTEM *Andropogon* spp											
147	fresh, early vegetative	2-00-821	27.0	18.0	0.80	0.69	0.41	0.24	0.41	—	—	—
148			100.0	68.0	3.00	2.58	1.52	0.91	1.55	—	—	—
149	fresh, mature	2-00-825	59.0	31.0	1.38	1.13	0.67	0.23	0.70	—	—	—
150			100.0	53.0	2.34	1.91	1.14	0.40	1.18	—	—	—
	BREWERS											
151	grains, dehy	5-02-141	92.0	81.0	3.57	2.46	2.10	1.25	1.47	2,293.0	3,056.0	1,969.0
152			100.0	88.0	3.88	2.67	2.28	1.36	1.60	2,491.0	3,319.0	2,139.0
153	grains, wet	5-02-142	21.0	18.5	0.81	0.56	0.48	0.29	0.37	—	—	—
154			100.0	88.0	3.88	2.67	2.28	1.36	1.60	—	—	—
	BROME *Bromus* spp											
155	fresh, early vegetative	2-00-892	34.0	25.0	1.11	0.97	0.58	0.37	0.57	—	—	—
156			100.0	74.0	3.26	2.85	1.70	1.08	1.69	—	—	—
157	fresh, mature	2-00-898	57.0	32.0	1.43	1.18	0.70	0.31	0.72	—	—	—
158			100.0	57.0	2.51	2.09	1.23	0.55	1.28	—	—	—

| | Horses | | | Swine | | | | Plant Cell Wall Constituents | | | | | | | |
Entry Number	TDN (%)	DE (Mcal/ kg)	ME (Mcal/ kg)	TDN (%)	DE (kcal/ kg)	ME (kcal/ kg)	Crude Protein (%)	Cell Walls (%)	Cellulose (%)	Hemicellulose (%)	Lignin (%)	Acid Detergent Fiber (%)	Crude Fiber (%)	Ether Extract (%)	Ash (%)
121	—	—	—	61.0	2,739.0	2,313.0	79.8	—	—	—	—	—	1.0	1.3	5.3
122	—	—	—	67.0	2,993.0	2,527.0	87.2	—	—	—	—	—	1.1	1.4	5.8
123	—	—	—	—	2,529.0	1,951.0	85.9	—	—	—	—	—	1.0	1.6	—
124	—	—	—	—	2,748.0	2,120.0	93.3	—	—	—	—	—	1.1	1.8	—
125	—	—	—	—	2,712.0	1,950.0	85.6	—	—	—	—	—	1.0	1.3	6.6
126	—	—	—	—	2,916.0	2,097.0	93.0	—	—	—	—	—	1.1	1.4	7.1
127	—	—	—	—	—	—	4.9	—	—	—	—	—	6.6	1.0	2.4
128	—	—	—	—	—	—	18.7	—	—	—	—	—	25.5	3.7	9.1
129	47.0	1.86	1.53	—	—	—	—	—	—	—	—	—	—	—	—
130	48.0	1.92	1.58	—	—	—	—	—	—	—	—	—	—	—	—
131	—	—	—	—	—	—	5.2	—	7.0	6.0	—	8.0	8.1	1.6	2.8
132	—	—	—	—	—	—	14.9	—	20.0	17.0	—	24.0	23.0	4.5	8.0
133	—	—	—	—	—	—	5.4	17.0	8.0	—	1.0	9.0	7.8	1.1	2.9
134	—	—	—	—	—	—	17.4	55.0	26.0	—	3.0	29.0	25.3	3.6	9.4
135	—	—	—	—	—	—	5.8	23.0	10.0	6.0	1.0	11.0	9.6	1.4	2.5
136	—	—	—	—	—	—	16.6	65.0	28.0	16.0	4.0	32.0	27.4	3.9	7.1
137	—	—	—	—	—	—	4.9	29.0	14.0	8.0	2.0	16.0	12.7	1.5	3.1
138	—	—	—	—	—	—	11.6	68.0	33.0	18.0	5.0	38.0	30.3	3.6	7.3
139	—	—	—	—	—	—	4.0	29.0	14.0	9.0	2.0	17.0	13.4	1.3	2.6
140	—	—	—	—	—	—	9.5	69.0	34.0	21.0	6.0	40.0	32.2	3.1	6.2
141	44.0	1.77	1.45	—	—	—	11.6	—	—	—	—	—	27.6	3.1	5.9
142	50.0	1.99	1.63	—	—	—	13.0	—	—	—	—	—	31.0	3.5	6.6
143	45.0	1.81	1.48	—	—	—	8.2	—	—	—	—	—	29.9	3.0	5.4
144	49.0	1.96	1.61	—	—	—	8.9	—	—	—	—	—	32.5	3.3	5.9
145	—	—	—	—	—	—	5.8	—	—	—	—	—	13.5	1.7	3.6
146	—	—	—	—	—	—	13.9	—	—	—	—	—	32.6	4.2	8.8
147	—	—	—	—	—	—	3.4	—	—	—	—	—	6.7	0.7	2.4
148	—	—	—	—	—	—	12.8	—	—	—	—	—	24.9	2.8	8.9
149	—	—	—	—	—	—	3.4	—	—	—	—	—	20.2	1.4	3.3
150	—	—	—	—	—	—	5.8	—	—	—	—	—	34.2	2.4	5.6
151	48.0	—	—	58.0	2,487.0	2,285.0	27.1	42.0	—	—	6.0	22.0	13.2	6.6	3.6
152	52.0	—	—	63.0	2,701.0	2,482.0	29.4	46.0	—	—	6.0	24.0	14.4	7.2	3.9
153	—	—	—	—	—	—	4.9	9.0	—	—	1.0	5.0	3.2	1.4	1.0
154	—	—	—	—	—	—	23.2	42.0	—	—	5.0	23.0	15.3	6.5	4.8
155	—	—	—	—	—	—	6.1	19.0	9.0	—	1.0	11.0	8.1	1.3	3.6
156	—	—	—	—	—	—	18.0	56.0	27.0	—	3.0	31.0	24.0	3.7	10.7
157	—	—	—	—	—	—	3.6	41.0	20.0	—	5.0	25.0	21.5	1.2	—
158	—	—	—	—	—	—	6.4	72.0	35.0	—	9.0	44.0	38.0	2.2	—

TABLE 1 Composition of Important Feeds: Energy Values, Proximate Analyses, Plant Cell Wall Constituents, and Acid Detergent Fiber—*Continued*

Entry Number	Feed Name Description	International Feed Number	Dry Matter (%)	Ruminants					Dairy Cattle	Chickens		
				TDN (%)	DE (Mcal/kg)	ME (Mcal/kg)	NE$_m$ (Mcal/kg)	NE$_g$ (Mcal/kg)	NE$_l$ (Mcal/kg)	ME$_n$ (kcal/kg)	TME (kcal/kg)	NE$_p$ (kcal/kg)
159	hay, sun-cured, late vege-	1-00-887	88.0	60.0	2.63	2.26	1.34	0.80	1.36	—	—	—
160	tative		100.0	68.0	3.00	2.58	1.52	0.91	1.55	—	—	—
161	hay, sun-cured, late bloom	1-00-888	89.0	52.0	2.31	1.93	1.14	0.55	1.18	—	—	—
162			100.0	59.0	2.60	2.18	1.28	0.62	1.33	—	—	—
	BROME, CHEATGRASS *Bromus tectorum*											
163	fresh, early vegetative	2-00-908	22.0	15.0	0.65	0.56	0.33	0.19	0.34	—	—	—
164			100.0	67.0	2.95	2.53	1.50	0.88	1.52	—	—	—
165	fresh, dough stage	2-00-910	35.0	20.0	0.88	0.73	0.43	0.19	0.45	—	—	—
166			100.0	57.0	2.51	2.09	1.23	0.55	1.28	—	—	—
	BROME, SMOOTH *Bromus inermis*											
167	fresh, early vegetative	2-00-956	30.0	22.0	0.95	0.83	0.49	0.31	0.49	—	—	—
168			100.0	73.0	3.22	2.80	1.67	1.06	1.67	—	—	—
169	fresh, mature	2-08-364	55.0	29.0	1.28	1.05	0.63	0.22	0.65	—	—	—
170			100.0	53.0	2.34	1.91	1.14	0.40	1.18	—	—	—
171	hay, sun-cured, midbloom	1-05-633	90.0	51.0	2.23	1.85	1.09	0.46	1.13	—	—	—
172			100.0	56.0	2.47	2.04	1.21	0.51	1.25	—	—	—
173	hay, sun-cured, mature	1-00-944	93.0	48.0	2.12	1.73	1.03	0.33	1.07	—	—	—
174			100.0	52.0	2.29	1.87	1.12	0.36	1.15	—	—	—
	BROOMCORN MILLET— SEE MILLET, PROSO											
	BUCKWHEAT, COMMON *Fagopyrum sagittatum*											
175	grain	4-00-994	88.0	63.0	2.79	2.42	1.44	0.90	1.44	2,647.0	2,703.0	1,801.0
176			100.0	72.0	3.17	2.76	1.64	1.03	1.64	3,016.0	3,081.0	2,053.0
177	middlings	5-00-991	89.0	75.0	3.30	2.93	1.78	1.20	1.73	—	—	—
178			100.0	84.0	3.70	3.29	2.00	1.35	1.94	—	—	—
	BUTTERMILK											
179	condensed (Cattle)	5-01-159	29.0	26.0	1.14	1.02	0.62	0.43	0.60	—	—	—
180			100.0	88.0	3.88	3.47	2.12	1.45	2.04	—	—	—
181	dehy (Cattle)	5-01-160	92.0	82.0	3.62	3.24	1.99	1.36	1.90	2,752.0	—	1,731.0
182			100.0	89.0	3.92	3.51	2.15	1.47	2.06	2,982.0	—	1,876.0
	CANARYGRASS, REED *Phalaris arundinacea*											
183	fresh	2-01-113	27.0	17.0	0.76	0.65	0.38	0.22	0.39	—	—	—
184			100.0	65.0	2.87	2.45	1.44	0.82	1.47	—	—	—
185	hay, sun-cured	1-01-104	91.0	50.0	2.21	1.82	1.08	0.43	1.12	—	—	—
186			100.0	55.0	2.43	2.00	1.19	0.47	1.23	—	—	—
	CARROT *Daucus* spp											
187	roots, fresh	4-01-145	12.0	10.0	0.44	0.39	0.24	0.16	0.23	471.0	—	442.0
188			100.0	84.0	3.70	3.29	2.00	1.35	1.94	3,979.0	—	3,737.0
	CASEIN											
189	dehy (cattle)	5-01-162	91.0	81.0	3.56	3.18	1.95	1.33	1.87	4,117.0	—	—
190			100.0	89.0	3.92	3.51	2.15	1.47	2.06	4,544.0	—	—
191	dehy, vitamin-free (cattle)	5-20-679	92.0	82.0	3.61	3.23	1.98	1.35	1.90	—	—	—
192			100.0	89.0	3.92	3.51	2.15	1.47	2.06	—	—	—

| | Horses | | | Swine | | | | Plant Cell Wall Constituents | | | | | | | |
Entry Number	TDN (%)	DE (Mcal/ kg)	ME (Mcal/ kg)	TDN (%)	DE (kcal/ kg)	ME (kcal/ kg)	Crude Protein (%)	Cell Walls (%)	Cellulose (%)	Hemicellulose (%)	Lignin (%)	Acid Detergent Fiber (%)	Crude Fiber (%)	Ether Extract (%)	Ash (%)
159	42.0	1.67	1.37	—	—	—	14.0	57.0	28.0	—	4.0	31.0	26.3	2.3	8.3
160	47.0	1.90	1.56	—	—	—	16.0	65.0	32.0	—	4.0	35.0	30.0	2.6	9.4
161	39.0	1.57	1.29	—	—	—	8.9	60.0	32.0	—	7.0	38.0	32.9	2.0	7.5
162	44.0	1.77	1.45	—	—	—	10.0	68.0	36.0	—	8.0	43.0	37.0	2.3	8.4
163	—	—	—	—	—	—	3.5	—	—	—	—	—	5.0	0.6	2.1
164	—	—	—	—	—	—	15.8	—	—	—	—	—	22.9	2.7	9.6
165	—	—	—	—	—	—	—	—	—	—	—	—	—	—	—
166	—	—	—	—	—	—	—	—	—	—	—	—	—	—	—
167	—	—	—	—	—	—	6.3	—	—	—	—	—	6.7	1.2	3.0
168	—	—	—	—	—	—	21.3	—	—	—	—	—	22.8	4.2	10.1
169	—	—	—	—	—	—	3.3	—	—	—	—	—	19.1	1.3	3.8
170	—	—	—	—	—	—	6.0	—	—	—	—	—	34.8	2.4	6.9
171	45.0	1.78	1.46	—	—	—	13.2	55.0	28.0	20.0	4.0	33.0	28.8	2.3	9.0
172	49.0	1.97	1.61	—	—	—	14.6	61.0	31.0	22.0	4.0	37.0	31.8	2.6	10.0
173	45.0	1.80	1.48	—	—	—	5.4	65.0	33.0	23.0	7.0	42.0	29.8	2.8	6.7
174	49.0	1.94	1.59	—	—	—	5.8	71.0	36.0	25.0	8.0	45.0	32.2	3.0	7.2
175	62.0	—	—	68.0	2,982.0	2,861.0	11.0	—	—	—	—	—	10.4	2.5	2.1
176	71.0	—	—	77.0	3,399.0	3,261.0	12.5	—	—	—	—	—	11.8	2.8	2.3
177	—	—	—	—	—	—	29.8	—	—	—	—	—	7.4	7.3	4.9
178	—	—	—	—	—	—	33.5	—	—	—	—	—	8.3	8.2	5.5
179	—	—	—	22.0	974.0	862.0	10.8	—	—	—	—	—	0.1	2.4	3.6
180	—	—	—	76.0	3,315.0	2,935.0	36.9	—	—	—	—	—	0.3	8.3	12.3
181	—	—	—	77.0	3,411.0	3,046.0	31.7	—	—	—	—	—	0.3	4.7	9.1
182	—	—	—	84.0	3,696.0	3,300.0	34.4	—	—	—	—	—	0.4	5.0	9.9
183	—	—	—	—	—	—	3.1	—	—	5.0	—	—	7.8	0.9	2.2
184	—	—	—	—	—	—	11.6	—	—	20.0	—	—	29.5	3.5	8.3
185	44.0	1.83	1.50	—	—	—	9.4	58.0	24.0	21.0	3.0	33.0	30.1	2.8	7.2
186	48.0	2.01	1.65	—	—	—	10.3	64.0	26.0	23.0	4.0	36.0	33.0	3.1	7.9
187	9.0	—	—	11.0	471.0	442.0	1.2	1.0	1.0	—	0.0	1.0	1.2	0.2	1.0
188	77.0	—	—	90.0	3,979.0	3,737.0	9.9	9.0	7.0	—	0.0	8.0	9.7	1.4	8.2
189	—	—	—	80.0	3,507.0	2,710.0	84.0	—	—	—	—	—	0.2	0.6	2.2
190	—	—	—	88.0	3,870.0	2,991.0	92.7	—	—	—	—	—	0.2	0.7	2.4
191	—	—	—	—	—	—	—	—	—	—	—	—	—	—	—
192	—	—	—	—	—	—	—	—	—	—	—	—	—	—	—

TABLE 1   Composition of Important Feeds: Energy Values, Proximate Analyses, Plant Cell Wall Constituents, and Acid Detergent Fiber—*Continued*

Entry Number	Feed Name Description	International Feed Number	Dry Matter (%)	Ruminants TDN (%)	DE (Mcal/kg)	ME (Mcal/kg)	NE_m (Mcal/kg)	NE_g (Mcal/kg)	Dairy Cattle NE_l (Mcal/kg)	Chickens ME_n (kcal/kg)	TME (kcal/kg)	NE_p (kcal/kg)
193	precipitated dehy	5-20-837	92.0	82.0	3.61	3.23	1.98	1.35	1.90	—	—	—
194	(cattle)		100.0	89.0	3.92	3.51	2.15	1.47	2.06	—	—	—
	CASSAVA, COMMON *Manihot esculenta*											
195	tubers, dehy	4-09-598	88.0	74.0	3.29	2.93	1.78	1.20	1.72	—	—	—
196			100.0	85.0	3.75	3.34	2.03	1.37	1.96	—	—	—
197	tubers, fresh	4-09-599	37.0	30.0	1.31	1.15	0.70	0.46	0.68	—	—	—
198			100.0	80.0	3.53	3.11	1.88	1.24	1.84	—	—	—
	CATTLE *Bos taurus* buttermilk—see Buttermilk											
199	lips, fresh	5-07-940	30.0	28.0	1.24	1.12	0.69	0.48	0.66	—	—	—
200			100.0	94.0	4.14	3.74	2.31	1.59	2.18	—	—	—
201	livers, fresh	5-01-166	28.0	27.0	1.18	1.07	0.66	0.46	0.62	—	—	—
202			100.0	96.0	4.23	3.83	2.37	1.64	2.23	—	—	—
203	manure, dehy, all forage	1-28-274	92.0	23.0	1.01	0.61	0.70	—	0.45	—	—	—
204			100.0	25.0	1.10	0.66	0.76	—	0.49	—	—	—
205	manure, dehy, feedlot	1-28-213	92.0	40.0	1.74	1.35	0.87	—	0.86	—	—	—
206	(High concentrate)		100.0	43.0	1.90	1.47	0.94	—	0.93	—	—	—
207	manure, dehy, forage and	1-28-214	92.0	28.0	1.22	0.82	0.72	—	0.57	—	—	—
208	concentrate		100.0	30.0	1.32	0.89	0.78	—	0.62	—	—	—
	milk—see Milk skim milk—see Milk											
209	spleens, fresh	5-07-942	24.0	23.0	1.01	0.91	0.56	0.39	0.53	—	—	—
210			100.0	95.0	4.19	3.78	2.34	1.61	2.21	—	—	—
211	udders, fresh	5-07-943	20.0	20.0	0.86	0.78	0.48	0.33	0.45	—	—	—
212			100.0	96.0	4.23	3.83	2.37	1.64	2.23	—	—	—
	CEREALS											
213	screenings	4-02-156	90.0	61.0	2.69	2.32	1.37	0.82	1.39	1,833.0	3,447.0	1,524.0
214			100.0	68.0	3.00	2.58	1.52	0.91	1.55	2,042.0	3,840.0	1,698.0
215	screenings refuse	4-02-151	91.0	54.0	2.40	2.02	1.19	0.59	1.23	—	—	—
216			100.0	60.0	2.65	2.22	1.31	0.65	1.35	—	—	—
217	screeenings uncleaned	4-02-153	92.0	60.0	2.64	2.25	1.33	0.75	1.35	—	—	—
218			100.0	65.0	2.87	2.45	1.44	0.82	1.47			
	CHICKEN *Gallus domesticus*											
219	hens, whole, fresh	5-07-950	33.0	31.0	1.37	1.23	0.76	0.52	0.72	1,048.0	—	768.0
220			100.0	94.0	4.14	3.74	2.31	1.59	2.18	3,175.0	—	2,327.0
	CITRUS *Citrus* spp											
221	pulp fines (Dried citrus	4-01-235	91.0	73.0	3.21	2.83	1.71	1.13	1.67	1,325.0	—	921.0
222	meal)		100.0	80.0	3.53	3.11	1.88	1.24	1.84	1,456.0	—	1,012.0
223	pulp, silage	3-01-234	21.0	18.0	0.81	0.73	0.44	0.30	0.43	—	—	—
224			100.0	88.0	3.88	3.47	2.12	1.45	2.04	—	—	—
225	pulp without fines, dehy	4-01-237	91.0	75.0	3.30	2.92	1.77	1.18	1.72	1,337.0	—	943.0
226	(Dried citrus pulp) syrup—see Molasses and syrup		100.0	82.0	3.62	3.20	1.94	1.30	1.89	1,467.0	—	1,035.0
	CLOVER, ALSIKE *Trifolium hybridum*											
227	fresh, early vegetative	2-01-314	19.0	12.0	0.55	0.47	0.28	0.16	0.28	—	—	—
228			100.0	66.0	2.91	2.49	1.47	0.85	1.50	—	—	—

	Horses			Swine				Plant Cell Wall Constituents							
Entry Num- ber	TDN (%)	DE (Mcal/ kg)	ME (Mcal/ kg)	TDN (%)	DE (kcal/ kg)	ME (kcal/ kg)	Crude Pro- tein (%)	Cell Walls (%)	Cell- ulose (%)	Hemi- cell- ulose (%)	Lig- nin (%)	Acid Deter- gent Fiber (%)	Crude Fiber (%)	Ether Ex- tract (%)	Ash (%)
193	—	—	—	—	—	—	85.0	—	—	—	—	—	—	—	—
194	—	—	—	—	—	—	92.4	—	—	—	—	—	—	—	—
195	—	—	—	78.0	3,457.0	3,334.0	2.3	—	—	—	—	—	4.6	0.7	2.9
196	—	—	—	89.0	3,939.0	3,800.0	2.6	—	—	—	—	—	5.2	0.8	3.3
197	—	—	—	—	—	—	1.3	—	—	—	—	—	1.7	0.4	1.4
198	—	—	—	—	—	—	3.6	—	—	—	—	—	4.6	1.0	3.9
199	—	—	—	—	—	—	18.0	—	—	—	—	—	—	7.0	—
200	—	—	—	—	—	—	60.0	—	—	—	—	—	—	23.3	—
201	—	—	—	—	—	—	19.5	—	—	—	—	—	0.2	5.1	1.4
202	—	—	—	—	—	—	69.6	—	—	—	—	—	0.6	18.3	4.9
203	—	—	—	—	—	—	15.6	63.0	—	—	25.0	42.0	—	—	—
204	—	—	—	—	—	—	17.0	69.0	—	—	27.0	46.0	—	—	—
205	—	—	—	—	—	—	23.0	29.0	—	—	5.0	24.0	—	—	18.2
206	—	—	—	—	—	—	25.0	32.0	—	—	5.0	26.0	—	—	19.8
207	—	—	—	—	—	—	15.6	53.0	—	—	7.0	31.0	—	—	—
208	—	—	—	—	—	—	17.0	58.0	—	—	7.0	34.0	—	—	—
209	—	—	—	—	—	—	16.5	—	—	—	—	—	1.0	3.9	1.5
210	—	—	—	—	—	—	68.7	—	—	—	—	—	4.0	16.1	6.0
211	—	—	—	—	—	—	11.9	—	—	—	—	—	0.3	6.1	1.5
212	—	—	—	—	—	—	58.6	—	—	—	—	—	1.2	30.0	7.4
213	—	—	—	55.0	3,331.0	3,107.0	12.1	—	––	—	—	—	12.0	3.7	5.4
214	—	—	—	61.0	3,710.0	3,461.0	13.4	—	—	—	—	—	13.4	4.1	6.0
215	—	—	—	47.0	2,069.0	1,927.0	12.8	—	—	—	—	—	17.0	4.5	8.9
216	—	—	—	52.0	2,276.0	2,120.0	14.1	—	—	—	—	—	18.7	4.9	9.8
217	—	—	—	49.0	2,163.0	2,011.0	13.9	—	—	—	—	—	17.1	5.4	8.6
218	—	—	—	53.0	2,352.0	2,186.0	15.1	—	—	—	—	—	18.6	5.9	9.3
219	—	—	—	24.0	1,101.0	933.0	19.9	—	—	—	—	—	0.5	7.9	1.0
220	—	—	—	74.0	3,337.0	2,829.0	60.3	—	—	—	—	—	1.5	24.1	3.1
221	—	—	—	76.0	3,333.0	3,181.0	6.5	—	—	—	—	—	13.1	3.3	6.3
222	—	—	—	83.0	3,662.0	3,496.0	7.1	—	—	—	—	—	14.4	3.6	6.9
223	—	—	—	—	—	—	1.5	—	—	—	—	—	3.3	2.0	1.2
224	—	—	—	—	—	—	7.3	—	—	—	—	—	15.6	9.7	5.5
225	—	—	—	46.0	2,286.0	2,163.0	6.1	21.0	—	—	3.0	20.0	11.6	3.4	6.0
226	—	—	—	50.0	2,508.0	2,374.0	6.7	23.0	—	—	3.0	22.0	12.7	3.7	6.6
227	—	—	—	—	—	—	4.5	—	—	—	—	—	3.3	0.6	2.4
228	—	—	—	—	—	—	24.1	—	—	—	—	—	17.5	3.2	12.8

TABLE 1   Composition of Important Feeds: Energy Values, Proximate Analyses, Plant Cell Wall Constituents, and Acid Detergent Fiber—*Continued*

Entry Number	Feed Name Description	International Feed Number	Dry Matter (%)	Ruminants TDN (%)	DE (Mcal/kg)	ME (Mcal/kg)	NE$_m$ (Mcal/kg)	NE$_g$ (Mcal/kg)	Dairy Cattle NE$_l$ (Mcal/kg)	Chickens ME$_n$ (kcal/kg)	TME (kcal/kg)	NE$_p$ (kcal/kg)
229	hay, sun-cured	1-01-313	88.0	51.0	2.25	1.87	1.11	0.51	1.14	—	—	—
230			100.0	58.0	2.56	2.13	1.26	0.58	1.30	—	—	—
	CLOVER, CRIMSON *Trifolium incarnatum*											
231	fresh, early vegetative	2-20-890	18.0	11.0	0.50	0.42	0.25	0.14	0.26	—	—	—
232			100.0	63.0	2.78	2.36	1.39	0.75	1.42	—	—	—
233	hay, sun-cured	1-01-328	87.0	50.0	2.19	1.82	1.08	0.48	1.11	—	—	—
234			100.0	57.0	2.51	2.09	1.23	0.55	1.28	—	—	—
	CLOVER, LADINO *Trifolium repens*											
235	fresh, early vegetative	2-01-380	19.0	13.0	0.58	0.50	0.30	0.18	0.30	—	—	—
236			100.0	68.0	3.00	2.58	1.52	0.91	1.55	—	—	—
237	hay, sun-cured	1-01-378	90.0	58.0	2.57	2.19	1.29	0.73	1.32	—	—	—
238			100.0	65.0	2.87	2.45	1.44	0.82	1.47	—	—	—
	CLOVER, RED *Trifolium pratense*											
239	fresh, regrowth early vege-	2-28-255	18.0	12.0	0.54	0.46	0.27	0.16	0.28	—	—	—
240	tative		100.0	68.0	3.00	2.58	1.52	0.91	1.55	—	—	—
241	fresh, early bloom	2-01-428	20.0	14.0	0.60	0.52	0.31	0.19	0.31	—	—	—
242			100.0	69.0	3.04	2.62	1.55	0.94	1.57	—	—	—
243	fresh, full bloom	2-01-429	26.0	17.0	0.74	0.63	0.37	0.21	0.38	—	—	—
244			100.0	64.0	2.82	2.40	1.41	0.78	1.45	—	—	—
245	hay, sun-cured	1-01-415	89.0	49.0	2.15	1.77	1.05	0.42	1.09	—	—	—
246			100.0	55.0	2.43	2.00	1.19	0.47	1.23	—	—	—
	COCONUT *Cocos nucifera*											
247	meats, meal mech extd	5-01-572	92.0	76.0	3.34	2.96	1.79	1.20	1.74	1,514.0	—	1,339.0
248	(Copra meal)		100.0	82.0	3.62	3.20	1.94	1.30	1.89	1,640.0	—	1,451.0
249	meats, meal solv extd	5-01-573	91.0	68.0	3.01	2.63	1.57	1.01	1.56	1,506.0	—	1,329.0
250	(Copra meal)		100.0	75.0	3.31	2.89	1.73	1.11	1.72	1,653.0	—	1,460.0
	COFFEE *Coffea* spp											
251	fruit with hulls without seeds,	1-09-648	89.0	55.0	2.43	2.05	1.21	0.64	1.24	—	—	—
252	dehy (Coffee pulp with hulls)		100.0	62.0	2.73	2.31	1.36	0.72	1.40	—	—	—
253	fruit without seeds, dehy	1-09-734	87.0	50.0	2.18	1.82	1.07	0.47	1.11	—	—	—
254	(Coffee pulp)		100.0	57.0	2.51	2.09	1.23	0.55	1.28	—	—	—
255	grounds, wet	1-01-576	74.0	38.0	1.69	1.38	0.82	0.26	0.85	—	—	—
256			100.0	52.0	2.29	1.87	1.12	0.36	1.15	—	—	—
257	hulls	1-01-577	90.0	0.0	0.0	0.0	0.0	0.0	0.0	—	—	—
258			100.0	0.0	0.0	0.0	0.0	0.0	0.0	—	—	—
	CORN, DENT YELLOW *Zea mays indentata*											
259	aerial part with ears, sun-	1-28-231	81.0	53.0	2.33	1.99	1.17	0.66	1.20	—	—	—
260	cured (Fodder)		100.0	65.0	2.87	2.45	1.44	0.82	1.47	—	—	—
261	aerial part with ears, sun-	1-28-232	82.0	56.0	2.50	2.15	1.27	0.77	1.29	—	—	—
262	cured, mature (Fodder)		100.0	69.0	3.04	2.62	1.55	0.94	1.57	—	—	—
263	aerial part without ears	1-28-233	85.0	51.0	2.21	1.85	1.09	0.53	1.13	—	—	—
264	without husks, sun-cured (Stover) (Straw)		100.0	59.0	2.60	2.18	1.28	0.62	1.33	—	—	—

	Horses			Swine				Plant Cell Wall Constituents							
Entry Number	TDN (%)	DE (Mcal/ kg)	ME (Mcal/ kg)	TDN (%)	DE (kcal/ kg)	ME (kcal/ kg)	Crude Protein (%)	Cell Walls (%)	Cell- ulose (%)	Hemi- cell- ulose (%)	Lig- nin (%)	Acid Deter- gent Fiber (%)	Crude Fiber (%)	Ether Ex- tract (%)	Ash (%)
229	42.0	1.70	1.39	—	—	—	13.1	—	—	11.0	—	—	26.5	2.7	7.7
230	48.0	1.93	1.59	—	—	—	14.9	—	—	13.0	—	—	30.1	3.0	8.7
231	—	—	—	—	—	—	3.1	—	—	—	—	—	5.0	—	—
232	—	—	—	—	—	—	17.0	—	—	—	—	—	28.0	—	—
233	45.0	1.80	1.47	—	—	—	16.1	—	—	—	—	—	26.3	2.1	9.6
234	52.0	2.06	1.69	—	—	—	18.4	—	—	—	—	—	30.1	2.4	11.0
235	—	—	—	—	—	—	5.3	—	—	—	—	—	2.7	0.5	2.6
236	—	—	—	—	—	—	27.2	—	—	—	—	—	14.0	2.5	13.5
237	45.0	1.81	1.49	—	—	—	19.7	32.0	—	—	6.0	29.0	19.1	2.4	9.0
238	51.0	2.02	1.66	—	—	—	22.0	36.0	—	—	7.0	32.0	21.2	2.7	10.1
239	—	—	—	—	—	—	3.8	—	—	—	—	—	—	—	—
240	—	—	—	—	—	—	21.0	—	—	—	—	—	—	—	—
241	—	—	—	—	—	—	3.8	—	—	—	—	—	4.6	1.0	2.0
242	—	—	—	—	—	—	19.4	—	—	—	—	—	23.2	5.0	10.2
243	—	—	—	—	—	—	3.8	—	—	—	—	—	6.8	0.8	2.0
244	—	—	—	—	—	—	14.6	—	—	—	—	—	26.1	2.9	7.8
245	44.0	1.76	1.45	—	—	—	14.2	50.0	23.0	8.0	9.0	—	25.5	2.4	7.5
246	50.0	1.99	1.63	—	—	—	16.0	56.0	26.0	9.0	10.0	—	28.8	2.8	8.5
247	—	—	—	75.0	3,739.0	3,420.0	20.7	—	—	—	—	—	11.8	6.4	6.8
248	—	—	—	82.0	4,051.0	3,706.0	22.4	—	—	—	—	—	12.8	6.9	7.3
249	—	—	—	73.0	3,216.0	3,053.0	21.3	—	—	—	—	—	14.0	3.5	6.0
250	—	—	—	80.0	3,532.0	3,353.0	23.4	—	—	—	—	—	15.4	3.9	6.6
251	43.0	1.72	1.41	67.0	2,942.0	2,766.0	8.7	—	—	—	—	—	22.1	2.8	4.1
252	48.0	1.94	1.59	75.0	3,317.0	3,118.0	9.8	—	—	—	—	—	24.9	3.2	4.6
253	42.0	1.68	1.38	—	—	—	11.8	59.0	29.0	4.0	25.0	56.0	27.0	2.5	6.5
254	48.0	1.93	1.58	—	—	—	13.6	68.0	33.0	4.0	29.0	64.0	31.1	2.9	7.4
255	—	—	—	—	—	—	0.0	55.0	—	—	7.0	50.0	21.5	9.3	1.2
256	—	—	—	—	—	—	0.0	74.0	—	—	9.0	68.0	29.2	12.6	1.6
257	—	—	—	—	—	—	—	84.0	45.0	—	18.0	64.0	32.6	7.4	4.9
258	—	—	—	—	—	—	—	93.0	50.0	—	20.0	71.0	36.2	8.2	5.4
259	37.0	1.48	1.22	—	—	—	7.2	45.0	23.0	—	2.0	27.0	20.5	2.0	5.5
260	45.0	1.83	1.50	—	—	—	8.9	55.0	28.0	—	3.0	33.0	25.2	2.4	6.8
261	35.0	1.43	1.17	—	—	—	6.6	—	—	—	—	—	18.6	1.9	4.4
262	43.0	1.74	1.43	—	—	—	8.0	—	—	—	—	—	22.6	2.3	5.4
263	32.0	1.31	1.07	—	—	—	5.6	57.0	21.0	—	9.0	33.0	29.3	1.1	6.1
264	37.0	1.53	1.26	—	—	—	6.6	67.0	25.0	—	11.0	39.0	34.4	1.3	7.2

TABLE 1   Composition of Important Feeds: Energy Values, Proximate Analyses, Plant Cell Wall Constituents, and Acid Detergent Fiber—*Continued*

Entry Number	Feed Name Description	International Feed Number	Dry Matter (%)	Ruminants					Dairy Cattle NE_l (Mcal/kg)	Chickens		
				TDN (%)	DE (Mcal/kg)	ME (Mcal/kg)	NE_m (Mcal/kg)	NE_g (Mcal/kg)		ME_n (kcal/kg)	TME (kcal/kg)	NE_p (kcal/kg)
265	cobs, ground	1-28-234	90.0	45.0	1.99	1.60	0.97	0.25	1.00	1,651.0	—	—
266			100.0	50.0	2.21	1.78	1.07	0.28	1.11	1,830.0	—	—
267	distillers grains, dehy	5-28-235	94.0	80.0	3.55	3.16	1.93	1.31	1.86	1,972.0	—	1,656.0
268			100.0	86.0	3.79	3.38	2.06	1.40	1.99	2,108.0	—	1,771.0
269	distillers grains with	5-28-236	92.0	80.0	3.56	3.19	1.95	1.33	1.87	2,535.0	2,970.0	1,951.0
270	solubles, dehy		100.0	88.0	3.88	3.47	2.12	1.45	2.04	2,760.0	3,234.0	2,124.0
271	distillers solubles, dehy	5-28-237	93.0	81.0	3.60	3.22	1.97	1.34	1.89	2,915.0	3,045.0	2,145.0
272			100.0	88.0	3.88	3.47	2.12	1.45	2.04	3,143.0	3,284.0	2,313.0
273	ears, ground (Corn and	4-28-238	87.0	72.0	3.17	2.81	1.70	1.14	1.66	2,730.0	—	1,968.0
274	cob meal)		100.0	83.0	3.66	3.25	1.97	1.32	1.91	3,155.0	—	2,274.0
275	ears with husks, silage	3-28-239	44.0	32.0	1.43	1.25	0.74	0.47	0.74	—	—	—
276			100.0	74.0	3.26	2.85	1.70	1.08	1.69	—	—	—
277	germs, meal wet milled	5-28-240	91.0	67.0	2.97	2.59	1.55	0.99	1.54	1,690.0	—	—
278	solv extd		100.0	74.0	3.26	2.85	1.70	1.08	1.69	1,854.0	—	—
279	gluten, meal	5-28-241	91.0	78.0	3.46	3.09	1.88	1.27	1.81	2,992.0	3,623.0	1,971.0
280			100.0	86.0	3.79	3.38	2.06	1.40	1.99	3,278.0	3,969.0	2,160.0
281	gluten, meal, 60% protein	5-28-242	90.0	81.0	3.54	3.17	1.95	1.33	1.86	3,689.0	4,003.0	2,724.0
282			100.0	89.0	3.92	3.51	2.15	1.47	2.06	4,086.0	4,434.0	3,017.0
283	gluten with bran (Corn	5-28-243	90.0	75.0	3.29	2.92	1.77	1.19	1.72	1,731.0	2,368.0	1,162.0
284	gluten feed)		100.0	83.0	3.66	3.25	1.97	1.32	1.91	1,924.0	2,632.0	1,291.0
285	grain	4-02-935	89.0	77.0	3.40	3.03	1.85	1.26	1.78	3,383.0	3,671.0	2,491.0
286			100.0	87.0	3.84	3.42	2.09	1.42	2.01	3,818.0	4,143.0	2,812.0
287	grain, boiled dehy	4-02-853	88.0	74.0	3.26	2.90	1.76	1.19	1.71	—	—	—
288			100.0	84.0	3.70	3.29	2.00	1.35	1.94	—	—	—
289	grain, cracked	4-20-698	89.0	71.0	3.14	2.77	1.67	1.11	1.64	—	—	—
290			100.0	80.0	3.53	3.11	1.88	1.24	1.84	—	—	—
291	grain, flaked	4-28-244	89.0	78.0	3.45	3.08	1.89	1.28	1.81	—	—	—
292			100.0	88.0	3.88	3.47	2.12	1.45	2.04	—	—	—
293	grain, ground	4-26-023	88.0	75.0	3.29	2.93	1.79	1.21	1.73	3,394.0	—	—
294			100.0	85.0	3.75	3.34	2.03	1.37	1.96	3,862.0	—	—
295	grain, high moisture	4-20-770	77.0	71.0	3.12	2.80	1.73	1.18	1.64	—	—	—
296			100.0	92.0	4.06	3.65	2.25	1.54	2.13	—	—	—
297	grain, opaque 2 (High	4-28-253	90.0	80.0	3.54	3.17	1.94	1.32	1.86	3,369.0	—	2,484.0
298	lysine)		100.0	89.0	3.92	3.51	2.15	1.47	2.06	3,738.0	—	2,756.0
299	grits (Hominy grits)	4-03-010	88.0	81.0	3.54	3.18	1.96	1.34	1.86	—	—	—
300			100.0	91.0	4.01	3.60	2.22	1.52	2.11	—	—	—
301	grits by-product (Hominy	4-03-011	90.0	85.0	3.74	3.37	2.09	1.43	1.97	2,896.0	—	1,879.0
302	feed)		100.0	94.0	4.14	3.74	2.31	1.59	2.18	3,208.0	—	2,082.0
303	silage	3-02-912	30.0	21.0	0.91	0.79	0.47	0.28	0.47	—	—	—
304			100.0	69.0	3.04	2.62	1.55	0.94	1.57	—	—	—
305	silage, aerial part without	3-28-251	31.0	17.0	0.74	0.61	0.36	0.15	0.38	—	—	—
306	ears without husks (Stalklage) (Stover)		100.0	55.0	2.43	2.00	1.19	0.47	1.23	—	—	—
307	silage, aerial part without	3-28-252	26.0	15.0	0.67	0.56	0.33	0.15	0.34	—	—	—
308	ears without husks, milk stage		100.0	58.0	2.56	2.13	1.26	0.58	1.30	—	—	—
309	silage, few ears	3-28-245	29.0	18.0	0.80	0.68	0.40	0.21	0.41	—	—	—
310			100.0	62.0	2.73	2.31	1.36	0.72	1.40	—	—	—
311	silage, well eared	3-28-250	33.0	23.0	1.03	0.89	0.53	0.32	0.53	—	—	—
312			100.0	70.0	3.09	2.67	1.58	0.97	1.60	—	—	—
313	silage, milk stage	3-08-402	22.0	14.0	0.64	0.54	0.32	0.18	0.33	—	—	—
314			100.0	65.0	2.87	2.45	1.44	0.82	1.47	—	—	—
315	silage, dough stage	3-28-246	26.0	18.0	0.81	0.70	0.41	0.25	0.42	—	—	—
316			100.0	69.0	3.04	2.62	1.55	0.94	1.57	—	—	—

| | Horses | | | Swine | | | | Plant Cell Wall Constituents | | | | | | | |
| | | | | | | | Crude Pro-tein (%) | Cell Walls (%) | Cell-ulose (%) | Hemi-cell-ulose (%) | Lig-nin (%) | Acid Deter-gent Fiber (%) | Crude Fiber (%) | Ether Ex-tract (%) | Ash (%) |
Entry Num-ber	TDN (%)	DE (Mcal/ kg)	ME (Mcal/ kg)	TDN (%)	DE (kcal/ kg)	ME (kcal/ kg)									
265	28.0	1.17	0.96	—	—	—	2.8	80.0	25.0	—	6.0	32.0	32.7	0.7	1.5
266	31.0	1.30	1.07	—	—	—	3.2	89.0	28.0	—	7.0	35.0	36.2	0.7	1.7
267	—	—	—	65.0	2,748.0	2,062.0	21.6	40.0	—	—	—	—	11.3	9.2	2.3
268	—	—	—	69.0	2,938.0	2,205.0	23.0	43.0	—	—	—	—	12.1	9.8	2.4
269	—	—	—	—	3,131.0	2,819.0	23.0	40.0	13.0	—	4.0	17.0	9.1	9.4	4.4
270	—	—	—	—	3,410.0	3,070.0	25.0	44.0	14.0	—	4.0	18.0	9.9	10.3	4.8
271	—	—	—	78.0	3,245.0	3,124.0	27.6	21.0	6.0	—	1.0	6.0	4.6	8.6	7.2
272	—	—	—	84.0	3,499.0	3,369.0	29.7	23.0	6.0	—	1.0	7.0	5.0	9.2	7.8
273	—	—	—	69.0	3,109.0	2,779.0	7.8	—	—	—	—	—	8.2	3.2	1.7
274	—	—	—	79.0	3,593.0	3,212.0	9.0	—	—	—	—	—	9.4	3.7	1.9
275	—	—	—	—	—	—	3.9	—	—	—	—	—	5.1	1.7	1.2
276	—	—	—	—	—	—	8.9	—	—	—	—	—	11.6	3.8	2.8
277	—	—	—	—	3,283.0	3,004.0	20.4	—	—	—	—	—	12.0	3.7	3.8
278	—	—	—	—	3,061.0	3,295.0	22.3	—	—	—	—	—	13.1	4.1	4.2
279	—	—	—	80.0	3,538.0	3,136.0	42.7	34.0	7.0	—	1.0	8.0	4.4	2.2	3.1
280	—	—	—	88.0	3,876.0	3,436.0	46.8	37.0	8.0	—	1.0	9.0	4.8	2.4	3.4
281	—	—	—	80.0	3,981.0	3,528.0	60.7	13.0	4.0	—	1.0	5.0	2.0	2.2	1.6
282	—	—	—	89.0	4,409.0	3,907.0	67.2	14.0	4.0	—	1.0	5.0	2.2	2.4	1.8
283	—	—	—	76.0	3,205.0	2,475.0	23.0	—	—	—	—	—	8.7	2.2	6.7
284	—	—	—	84.0	3,562.0	2,751.0	25.6	—	—	—	—	—	9.7	2.4	7.5
285	—	—	—	80.0	3,399.0	3,300.0	9.6	8.0	2.0	5.0	1.0	3.0	2.6	3.8	1.3
286	—	—	—	90.0	3,837.0	3,724.0	10.9	9.0	2.0	6.0	1.0	3.0	2.9	4.3	1.5
287	—	—	—	80.0	3,533.0	3,316.0	9.3	—	—	—	—	—	1.6	4.6	1.9
288	—	—	—	91.0	4,008.0	3,762.0	10.5	—	—	—	—	—	1.8	5.2	2.1
289	—	—	—	—	—	—	8.9	8.0	2.0	5.0	1.0	3.0	2.2	—	—
290	—	—	—	—	—	—	10.0	9.0	2.0	6.0	1.0	3.0	2.5	—	—
291	—	—	—	85.0	3,735.0	3,501.0	9.9	—	—	—	—	—	0.6	2.0	0.9
292	—	—	—	95.0	4,206.0	3,943.0	11.2	—	—	—	—	—	0.7	2.2	1.0
293	—	—	—	75.0	3,373.0	3,264.0	8.8	8.0	2.0	5.0	1.0	3.0	2.2	3.8	1.4
294	—	—	—	85.0	3,837.0	3,714.0	10.0	9.0	2.0	6.0	1.0	3.0	2.5	4.3	1.5
295	—	—	—	66.0	2,891.0	2,713.0	8.2	—	—	—	—	4.0	2.0	3.3	1.2
296	—	—	—	85.0	3,765.0	3,534.0	10.7	—	—	—	—	5.0	2.6	4.3	1.6
297	—	—	—	77.0	3,664.0	3,434.0	10.1	—	—	—	—	—	3.3	4.3	1.6
298	—	—	—	85.0	4,065.0	3,810.0	11.3	—	—	—	—	—	3.7	4.8	1.8
299	—	—	—	79.0	3,466.0	3,260.0	8.5	—	—	—	—	—	0.5	0.7	0.4
300	—	—	—	89.0	3,927.0	3,694.0	9.6	—	—	—	—	—	0.6	0.8	0.5
301	—	—	—	82.0	3,625.0	3,382.0	10.4	50.0	9.0	—	2.0	12.0	6.0	6.9	2.8
302	—	—	—	91.0	4,017.0	3,748.0	11.5	55.0	10.0	—	2.0	13.0	6.7	7.7	3.1
303	—	—	—	22.0	950.0	896.0	2.5	—	9.0	—	1.0	9.0	7.5	1.0	1.7
304	—	—	—	72.0	3,161.0	2,981.0	8.3	—	29.0	—	5.0	30.0	25.1	3.3	5.5
305	—	—	—	—	—	—	1.9	21.0	8.0	—	2.0	17.0	9.6	0.7	3.5
306	—	—	—	—	—	—	6.3	68.0	25.0	—	7.0	55.0	31.3	2.1	11.6
307	—	—	—	—	—	—	—	—	—	—	—	—	—	—	—
308	—	—	—	—	—	—	—	—	—	—	—	—	—	—	—
309	—	—	—	—	—	—	2.4	—	—	—	—	—	9.4	0.9	2.1
310	—	—	—	—	—	—	8.4	—	—	—	—	—	32.3	3.0	7.2
311	—	—	—	—	—	—	2.7	17.0	—	—	—	9.0	7.9	1.0	1.5
312	—	—	—	—	—	—	8.1	51.0	—	—	—	28.0	23.7	3.1	4.5
313	—	—	—	—	—	—	2.0	—	—	—	—	—	6.8	0.7	1.2
314	—	—	—	—	—	—	8.9	—	—	—	—	—	30.5	3.1	5.2
315	—	—	—	—	—	—	2.1	—	—	—	—	—	6.5	0.8	1.2
316	—	—	—	—	—	—	7.8	—	—	—	—	—	24.5	2.9	4.7

TABLE 1   Composition of Important Feeds: Energy Values, Proximate Analyses, Plant Cell Wall Constituents, and Acid Detergent Fiber—*Continued*

Entry Number	Feed Name Description	International Feed Number	Dry Matter (%)	Ruminants TDN (%)	DE (Mcal/ kg)	ME (Mcal/ kg)	NE$_m$ (Mcal/ kg)	NE$_g$ (Mcal/ kg)	Dairy Cattle NE$_l$ (Mcal/ kg)	Chickens ME$_n$ (kcal/ kg)	TME (kcal/ kg)	NE$_p$ (kcal/ kg)
	CORN, FLINT *Zea mays indurata*											
317	grain	4-02-948	89.0	83.0	3.67	3.31	2.05	1.41	1.93	—	—	—
318			100.0	94.0	4.14	3.74	2.31	1.59	2.18	—	—	—
	CORN, SWEET *Zea mays saccharata*											
319	process residue, fresh	2-02-975	77.0	54.0	2.37	2.05	1.21	0.74	1.23	—	—	—
320			100.0	70.0	3.09	2.67	1.58	0.97	1.60	—	—	—
321	process residue, silage	3-07-955	32.0	23.0	1.01	0.87	0.52	0.33	0.52	—	—	—
322			100.0	72.0	3.17	2.76	1.64	1.03	1.64	—	—	—
	COTTON *Gossypium* spp											
323	bolls, sun-cured	1-01-596	92.0	41.0	1.78	1.38	0.88	0.01	0.88	—	—	—
324			100.0	44.0	1.94	1.51	0.96	0.01	0.96	—	—	—
325	hulls	1-01-599	91.0	41.0	1.80	1.41	0.88	0.05	0.89	—	—	—
326			100.0	45.0	1.98	1.55	0.98	0.06	0.98	—	—	—
327	seeds, with lint	5-01-614	92.0	88.0	3.89	3.52	2.18	1.51	2.05	—	—	—
328			100.0	96.0	4.23	3.83	2.37	1.64	2.23	—	—	—
329	seeds, meal mech extd	5-01-609	93.0	55.0	2.46	2.06	1.22	0.61	1.25	—	—	—
330	(Whole pressed cottonseed)		100.0	60.0	2.65	2.22	1.31	0.65	1.35	—	—	—
331	seeds, meal mech extd,	5-01-625	92.0	67.0	2.97	2.58	1.54	0.97	1.54	1,970.0	—	1,299.0
332	36% protein		100.0	73.0	3.22	2.80	1.67	1.06	1.67	2,137.0	—	1,408.0
333	seeds, meal mech extd,	5-01-617	93.0	72.0	3.19	2.80	1.68	1.10	1.66	2,258.0	—	1,590.0
334	41% protein		100.0	78.0	3.44	3.02	1.82	1.19	1.79	2,437.0	—	1,716.0
335	seeds, meal prepressed	5-07-872	91.0	72.0	3.19	2.82	1.70	1.13	1.67	2,144.0	—	1,286.0
336	solv extd, 41% protein		100.0	80.0	3.53	3.11	1.88	1.24	1.84	2,368.0	—	1,420.0
337	seeds, meal prepressed	5-07-873	91.0	73.0	3.22	2.84	1.72	1.14	1.68	1,857.0	—	1,253.0
338	solv extd, 44% protein		100.0	80.0	3.53	3.11	1.88	1.24	1.84	2,033.0	—	1,372.0
339	seeds, meal solv extd, low	5-01-633	93.0	66.0	2.90	2.51	1.49	0.98	1.50	—	—	—
340	gossypol		100.0	71.0	3.13	2.71	1.61	1.00	1.62	—	—	—
341	seeds, meal solv extd, 41%	5-01-621	91.0	70.0	3.06	2.68	1.60	1.04	1.59	1,943.0	—	1,410.0
342	protein		100.0	76.0	3.35	2.93	1.76	1.14	1.74	2,131.0	—	1,546.0
343	seeds without hulls, meal	5-07-874	93.0	70.0	3.08	2.69	1.61	1.03	1.60	2,141.0	—	1,636.0
344	prepressed solv extd, 50% protein		100.0	75.0	3.31	2.89	1.73	1.11	1.72	2,302.0	—	1,759.0
	COWPEA, COMMON *Vigna sinensis*											
345	hay, sun-cured	1-01-645	90.0	53.0	2.34	1.96	1.15	0.56	1.19	—	—	—
346			100.0	59.0	2.60	2.18	1.28	0.62	1.33	—	—	—
	CRAB *Callinectes sapidus– Cancer* spp											
347	process residue, meal	5-01-663	92.0	27.0	1.18	0.78	0.72	—	0.55	1,825.0	—	1,316.0
348	(Crab meal)		100.0	29.0	1.28	0.84	0.78	—	0.59	1,977.0	—	1,425.0
	DISTILLERS GRAINS—SEE CORN, SEE SORGHUM											
	DROPSEED, SAND *Sporo- bolus cryptandrus*											
349	fresh, stem-cured	2-05-596	88.0	52.0	2.29	1.92	1.13	0.54	1.17	—	—	—
350			100.0	59.0	2.60	2.18	1.28	0.62	1.33	—	—	—

	Horses			Swine				Plant Cell Wall Constituents							
Entry Num- ber	TDN (%)	DE (Mcal/ kg)	ME (Mcal/ kg)	TDN (%)	DE (kcal/ kg)	ME (kcal/ kg)	Crude Pro- tein (%)	Cell Walls (%)	Cell- ulose (%)	Hemi- cell- ulose (%)	Lig- nin (%)	Acid Deter- gent Fiber (%)	Crude Fiber (%)	Ether Ex- tract (%)	Ash (%)
317	—	—	—	75.0	3,328.0	3,120.0	9.9	—	—	—	—	—	1.9	4.3	1.5
318	—	—	—	85.0	3,761.0	3,526.0	11.1	—	—	—	—	—	2.1	4.9	1.7
319	—	—	—	—	—	—	6.8	—	—	—	—	—	17.1	1.8	2.5
320	—	—	—	—	—	—	8.8	—	—	—	—	—	22.3	2.3	3.3
321	—	—	—	—	—	—	2.4	—	—	—	—	—	11.2	1.6	1.6
322	—	—	—	—	—	—	7.7	—	—	—	—	—	35.5	5.2	4.9
323	44.0	1.75	1.44	—	—	—	10.1	—	—	—	—	—	29.5	2.4	7.1
324	48.0	1.91	1.57	—	—	—	11.0	—	—	—	—	—	32.2	2.7	7.7
325	29.0	1.22	1.00	—	—	—	3.7	82.0	53.0	—	22.0	66.0	43.3	1.5	2.6
326	32.0	1.35	1.11	—	—	—	4.1	90.0	59.0	—	24.0	73.0	47.8	1.7	2.8
327	—	—	—	—	—	—	22.0	36.0	—	—	—	27.0	19.1	21.3	4.4
328	—	—	—	—	—	—	23.9	39.0	—	—	—	29.0	20.8	23.1	4.8
329	—	—	—	—	3,559.0	3,054.0	37.9	—	—	—	—	—	13.3	5.0	6.3
330	—	—	—	—	3,831.0	3,287.0	40.8	—	—	—	—	—	14.3	5.4	6.8
331	—	—	—	67.0	2,961.0	2,610.0	38.6	—	—	—	—	—	14.3	4.2	6.7
332	—	—	—	73.0	3,212.0	2,831.0	41.9	—	—	—	—	—	15.5	4.6	7.3
333	—	—	—	69.0	2,939.0	2,714.0	41.0	26.0	12.0	—	6.0	19.0	11.9	4.6	6.1
334	—	—	—	75.0	3,173.0	2,929.0	44.3	28.0	13.0	—	6.0	20.0	12.8	5.0	6.6
335	—	—	—	61.0	2,615.0	2,485.0	41.3	24.0	11.0	—	5.0	17.0	12.8	1.1	6.4
336	—	—	—	68.0	2,888.0	2,745.0	45.6	26.0	12.0	—	6.0	19.0	14.1	1.3	7.0
337	—	—	—	—	—	—	44.7	26.0	12.0	—	6.0	19.0	11.1	1.6	6.1
338	—	—	—	—	—	—	48.9	28.0	13.0	—	7.0	21.0	12.1	1.7	6.7
339	—	—	—	—	—	—	41.5	—	—	—	—	—	12.7	1.2	5.8
340	—	—	—	—	—	—	44.8	—	—	—	—	—	13.7	1.3	6.3
341	—	—	—	61.0	2,675.0	2,364.0	41.2	—	—	—	—	—	12.1	1.4	6.5
342	—	—	—	67.0	2,933.0	2,592.0	45.2	—	—	—	—	—	13.3	1.6	7.1
343	—	—	—	—	—	—	50.3	—	—	—	—	—	8.2	1.3	6.6
344	—	—	—	—	—	—	54.0	—	—	—	—	—	8.8	1.4	7.1
345	43.0	1.73	1.42	—	—	—	17.5	—	—	—	—	—	24.0	2.8	10.2
346	48.0	1.92	1.58	—	—	—	19.4	—	—	—	—	—	26.7	3.1	11.3
347	—	—	—	—	1,378.0	1,226.0	32.1	—	—	—	—	—	10.7	2.0	41.2
348	—	—	—	—	1,493.0	1,328.0	34.8	—	—	—	—	—	11.6	2.1	44.6
349	—	—	—	—	—	—	4.4	—	—	—	5.0	—	—	1.2	5.6
350	—	—	—	—	—	—	5.0	—	—	—	6.0	—	—	1.4	6.3

TABLE 1 Composition of Important Feeds: Energy Values, Proximate Analyses, Plant Cell Wall Constituents, and Acid Detergent Fiber—*Continued*

Entry Number	Feed Name Description	International Feed Number	Dry Matter (%)	Ruminants TDN (%)	DE (Mcal/ kg)	ME (Mcal/ kg)	NE_m (Mcal/ kg)	NE_g (Mcal/ kg)	Dairy Cattle NE_l (Mcal/ kg)	Chickens ME_n (kcal/ kg)	TME (kcal/ kg)	NE_p (kcal/ kg)
	EMMER *Triticum dicoccum*											
351	grain	4-01-830	91.0	72.0	3.20	2.83	1.71	1.13	1.67	—	—	—
352			100.0	80.0	3.53	3.11	1.88	1.24	1.84	—	—	—
	FATS AND OILS											
353	fat, animal, hydrolyzed	4-00-376	99.0	223.0	9.84	9.49	6.55	4.19	5.35	8,164.0	—	5,317.0
354			100.0	225.0	9.92	9.57	6.61	4.23	5.39	8,232.0	—	5,362.0
355	fat, animal, poultry	4-00-409	99.0	188.0	8.31	7.94	5.04	3.89	4.50	7,680.0	—	6,197.0
356			100.0	190.0	8.38	8.01	5.08	3.92	4.54	7,745.0	—	6,250.0
357	fat, swine (Lard)	4-04-790	99.0	189.0	8.31	7.95	5.05	3.89	4.50	8,658.0	9,161.0	—
358			100.0	190.0	8.38	8.01	5.08	3.92	4.54	8,723.0	9,230.0	—
359	oil, soybean	4-07-983	99.0	194.0	8.54	8.18	5.22	3.99	4.63	8,609.0	9,457.0	—
360			100.0	195.0	8.60	8.23	5.25	4.02	4.66	8,667.0	9,520.0	—
361	oil, vegetable	4-05-077	100.0	195.0	8.58	8.21	5.24	4.01	4.65	8,112.0	—	6,012.0
362			100.0	195.0	8.60	8.23	5.25	4.02	4.66	8,132.0	—	6,027.0
	FESCUE *Festuca* spp											
363	hay, sun-cured, early vege-	1-06-132	91.0	56.0	2.45	2.06	1.21	0.62	1.25	—	—	—
364	tative (South)		100.0	61.0	2.69	2.27	1.33	0.69	1.38	—	—	—
365	hay, sun-cured, late vege-	1-13-582	91.0	53.0	2.33	1.94	1.15	0.53	1.18	—	—	—
366	tative (South)		100.0	58.0	2.56	2.13	1.26	0.58	1.30	—	—	—
367	hay, sun-cured, early bloom	1-01-871	92.0	44.0	1.95	1.55	0.95	0.18	0.97	—	—	—
368	(South)		100.0	48.0	2.12	1.69	1.03	0.19	1.06	—	—	—
	FESCUE, ALTA *Festuca arundinacea*											
369	hay, sun-cured	1-05-684	91.0	57.0	2.49	2.10	1.24	0.65	1.27	—	—	—
370			100.0	62.0	2.73	2.31	1.36	0.72	1.40	—	—	—
	FESCUE, KENTUCKY 31 *Festuca arundinacea*											
371	fresh, vegetative	2-01-902	29.0	19.0	0.85	0.72	0.43	0.25	0.44	—	—	—
372			100.0	67.0	2.91	2.49	1.47	0.85	1.50	—	—	—
373	hay, sun-cured, early bloom	1-09-186	91.0	58.0	2.57	2.18	1.29	0.71	1.32	—	—	—
374			100.0	64.0	2.82	2.40	1.41	0.78	1.45	—	—	—
375	hay, sun-cured, midbloom	1-09-187	92.0	55.0	2.44	2.05	1.21	0.60	1.25	—	—	—
376			100.0	60.0	2.65	2.22	1.31	0.65	1.35	—	—	—
377	hay, sun-cured, full bloom	1-09-188	92.0	53.0	2.34	1.95	1.15	0.53	1.19	—	—	—
378			100.0	58.0	2.56	2.13	1.26	0.58	1.30	—	—	—
379	hay, sun-cured, mature	1-09-189	90.0	50.0	2.22	1.84	1.09	0.46	1.13	—	—	—
380			100.0	56.0	2.47	2.04	1.21	0.51	1.25	—	—	—
	FESCUE, MEADOW *Festuca elatior*											
381	fresh, vegetative	2-01-920	28.0	17.0	0.75	0.63	0.37	0.19	0.38	—	—	—
382			100.0	61.0	2.69	2.27	1.33	0.69	1.38	—	—	—
383	hay, sun-cured	1-01-912	88.0	52.0	2.29	1.91	1.13	0.54	1.17	—	—	—
384			100.0	59.0	2.60	2.18	1.28	.062	1.33	—	—	—
	FISH											
385	livers, meal mech extd	5-01-968	93.0	97.0	4.26	3.88	2.43	1.69	2.25	—	—	—
386			100.0	104.0	4.59	4.18	2.62	1.82	2.43	—	—	—
387	solubles, condensed	5-01-969	50.0	42.0	1.86	1.65	1.00	0.68	0.97	1,786.0	—	1,048.0
388			100.0	84.0	3.70	3.29	2.00	1.35	1.94	3,562.0	—	2,091.0

	Horses			Swine				Plant Cell Wall Constituents							
Entry Number	TDN (%)	DE (Mcal/ kg)	ME (Mcal/ kg)	TDN (%)	DE (kcal/ kg)	ME (kcal/ kg)	Crude Pro-tein (%)	Cell Walls (%)	Cell-ulose (%)	Hemi-cell-ulose (%)	Lig-nin (%)	Acid Deter-gent Fiber (%)	Crude Fiber (%)	Ether Ex-tract (%)	Ash (%)
351	—	—	—	70.0	3,088.0	2,884.0	11.7	—	—	—	—	—	9.6	2.0	3.5
352	—	—	—	77.0	3,401.0	3,176.0	12.9	—	—	—	—	—	10.6	2.2	3.9
353	—	—	—	209.0	9,272.0	8,274.0	—	—	—	—	—	—	—	98.7	—
354	—	—	—	211.0	9,350.0	8,343.0	—	—	—	—	—	—	—	99.5	—
355	—	—	—	196.0	8,635.0	7,976.0	—	—	—	—	—	—	—	99.1	—
356	—	—	—	198.0	8,708.0	8,044.0	—	—	—	—	—	—	—	100.0	—
357	—	—	—	—	7,986.0	7,791.0	—	—	—	—	—	—	—	99.6	—
358	—	—	—	—	8,046.0	7,850.0	—	—	—	—	—	—	—	100.3	—
359	—	—	—	—	7,512.0	7,234.0	1.4	—	—	—	—	—	—	95.4	0.3
360	—	—	—	—	7,562.0	7,283.0	1.4	—	—	—	—	—	—	96.0	0.3
361	—	—	—	206.0	8,841.0	8,027.0	—	—	—	—	—	—	—	99.7	—
362	—	—	—	206.0	8,863.0	8,047.0	—	—	—	—	—	—	—	99.9	—
363	—	—	—	—	—	—	11.3	52.0	25.0	—	3.0	29.0	23.7	3.1	10.9
364	—	—	—	—	—	—	12.4	57.0	28.0	—	3.0	32.0	26.0	3.4	12.0
365	—	—	—	—	—	—	9.6	58.0	28.0	—	4.0	33.0	30.0	2.7	9.6
366	—	—	—	—	—	—	10.5	64.0	31.0	—	4.0	36.0	33.0	3.0	10.5
367	—	—	—	—	—	—	8.7	66.0	30.0	—	5.0	36.0	34.0	1.8	9.2
368	—	—	—	—	—	—	9.5	72.0	33.0	—	5.0	39.0	37.0	2.0	10.0
369	41.0	1.75	1.44	—	—	—	9.3	64.0	—	24.0	—	37.0	32.5	2.0	7.0
370	45.0	1.92	1.58	—	—	—	10.2	70.0	—	26.0	—	41.0	35.7	2.2	7.7
371	—	—	—	—	—	—	4.2	—	—	—	—	—	7.1	1.6	2.9
372	—	—	—	—	—	—	14.5	—	—	—	—	—	24.6	5.5	9.9
373	0.0	0.18	0.14	—	—	—	18.4	54.0	26.0	—	3.0	29.0	21.5	6.0	8.9
374	1.0	0.19	0.16	—	—	—	20.2	59.0	29.0	—	3.0	32.0	23.6	6.6	9.8
375	23.0	0.99	0.81	—	—	—	15.2	58.0	28.0	—	3.0	32.0	23.6	5.7	8.4
376	25.0	1.07	0.88	—	—	—	16.4	63.0	30.0	—	4.0	35.0	25.5	6.1	9.1
377	42.0	1.70	1.40	—	—	—	11.1	61.0	29.0	—	5.0	36.0	25.1	4.9	7.2
378	46.0	1.86	1.52	—	—	—	12.1	67.0	32.0	—	5.0	39.0	27.4	5.3	7.9
379	47.0	1.85	1.52	—	—	—	8.3	63.0	31.0	—	6.0	38.0	29.3	3.8	5.8
380	52.0	2.06	1.69	—	—	—	9.2	70.0	34.0	—	7.0	42.0	32.6	4.3	6.4
381	—	—	—	—	—	—	3.7	—	—	—	—	—	8.2	1.2	2.3
382	—	—	—	—	—	—	13.5	—	—	—	—	—	29.7	4.2	8.2
383	41.0	1.63	1.34	—	—	—	8.0	57.0	33.0	—	6.0	38.0	29.1	2.1	7.2
384	46.0	1.86	1.53	—	—	—	9.1	65.0	38.0	—	7.0	43.0	33.1	2.4	8.2
385	—	—	—	—	—	—	62.8	—	—	—	—	—	1.2	17.3	6.1
386	—	—	—	—	—	—	67.7	—	—	—	—	—	1.3	18.6	6.6
387	—	—	—	44.0	1,898.0	1,613.0	32.7	—	—	—	—	—	0.5	5.6	9.6
388	—	—	—	88.0	3,784.0	3,217.0	65.3	—	—	—	—	—	0.9	11.2	19.2

TABLE 1   Composition of Important Feeds: Energy Values, Proximate Analyses, Plant Cell Wall Constituents, and Acid Detergent Fiber—*Continued*

Entry Number	Feed Name Description	International Feed Number	Dry Matter (%)	Ruminants TDN (%)	DE (Mcal/kg)	ME (Mcal/kg)	NE_m (Mcal/kg)	NE_g (Mcal/kg)	Dairy Cattle NE_l (Mcal/kg)	Chickens ME_n (kcal/kg)	TME (kcal/kg)	NE_p (kcal/kg)
389	solubles, dehy	5-01-971	93.0	77.0	3.39	3.01	1.83	1.22	1.77	2,911.0	—	2,003.0
390			100.0	83.0	3.66	3.25	1.97	1.32	1.91	3,140.0	—	2,161.0
	FISH, ALEWIFE *Pomolobus pseudoharengus*											
391	meal mech extd	5-09-830	90.0	69.0	3.05	2.68	1.61	1.05	1.59	3,494.0	—	2,725.0
392			100.0	77.0	3.40	2.98	1.79	1.17	1.77	3,889.0	—	3,033.0
	FISH, ANCHOVY *Engraulis ringen*											
393	meal mech extd	5-01-985	92.0	72.0	3.20	2.82	1.70	1.12	1.67	2,708.0	2,972.0	1,908.0
394			100.0	79.0	3.48	3.07	1.85	1.22	1.82	2,948.0	3,235.0	2,077.0
	FISH, HERRING *Clupea harengus*											
395	meal mech extd	5-02-000	92.0	76.0	3.36	2.98	1.81	1.21	1.76	3,261.0	3,377.0	2,101.0
396			100.0	83.0	3.66	3.25	1.97	1.32	1.91	3,548.0	3,674.0	2,285.0
	FISH, MENHADEN *Brevoortia tyrannus*											
397	meal mech extd	5-02-009	92.0	67.0	2.95	2.57	1.53	0.97	1.53	2,849.0	2,744.0	2,037.0
398			100.0	73.0	3.22	2.80	1.67	1.06	1.67	3,110.0	2,995.0	2,225.0
	FISH, REDFISH *Sciaenops ocellata*											
399	meal mech extd	5-07-973	93.0	72.0	3.16	2.78	1.66	1.09	1.65	3,229.0	2,359.0	1,908.0
400			100.0	77.0	3.40	2.98	1.79	1.17	1.77	3,467.0	2,533.0	2,048.0
	FISH, SALMON *Oncorhynchus* spp–*Salmo* spp											
401	meal mech extd	5-02-012	93.0	71.0	3.16	2.77	1.66	1.08	1.64	—	—	—
402			100.0	77.0	3.40	2.98	1.79	1.17	1.77	—	—	—
	FISH, SARDINE *Clupea* spp–*Sardinops* spp											
403	meal mech extd	5-02-015	93.0	70.0	3.08	2.69	1.61	1.03	1.60	2,896.0	—	2,005.0
404			100.0	75.0	3.31	2.89	1.73	1.11	1.72	3,109.0	—	2,152.0
	FISH, TUNA *Thunnus thynnus–Thunnus albacares*											
405	meal mech extd	5-02-023	93.0	65.0	2.86	2.47	1.47	0.90	1.48	2,813.0	—	1,979.0
406			100.0	70.0	3.09	2.67	1.58	0.97	1.60	3,032.0	—	2,134.0
	FISH, WHITE *Gadidae* (family)–*Lophiidae* (family)											
407	meal mech extd	5-02-025	91.0	70.0	3.10	2.72	1.63	1.06	1.61	2,593.0	2,387.0	1,821.0
408			100.0	77.0	3.40	2.98	1.79	1.17	1.77	2,843.0	2,616.0	1,997.0
	FLAX *Linum usitatissimum*											
409	seed screenings	4-02-056	91.0	58.0	2.57	2.19	1.29	0.72	1.32	—	—	—
410			100.0	64.0	2.82	2.40	1.41	0.78	1.45	—	—	—
411	seeds, meal mech extd	5-02-045	91.0	74.0	3.28	2.90	1.76	1.18	1.71	1,518.0	—	1,123.0
412	(Linseed meal)		100.0	82.0	3.62	3.20	1.94	1.30	1.89	1,673.0	—	1,237.0
413	seeds, meal solv extd	5-02-048	90.0	70.0	3.10	2.73	1.64	1.08	1.62	1,411.0	2,644.0	991.0
414	(Linseed meal)		100.0	78.0	3.44	3.02	1.82	1.19	1.79	1,565.0	2,931.0	1,099.0

	Horses			Swine			Crude Pro-tein (%)	Plant Cell Wall Constituents							
Entry Num-ber	TDN (%)	DE (Mcal/ kg)	ME (Mcal/ kg)	TDN (%)	DE (kcal/ kg)	ME (kcal/ kg)		Cell Walls (%)	Cell-ulose (%)	Hemi-cell-ulose (%)	Lig-nin (%)	Acid Deter-gent Fiber (%)	Crude Fiber (%)	Ether Ex-tract (%)	Ash (%)
389	—	—	—	66.0	2,922.0	2,814.0	64.1	—	—	—	—	—	1.4	8.2	12.5
390	—	—	—	72.0	3,153.0	3,036.0	69.2	—	—	—	—	—	1.5	8.9	13.5
391	—	—	—	68.0	3,798.0	3,334.0	36.4	—	—	—	—	—	1.4	9.8	16.7
392	—	—	—	76.0	4,227.0	3,711.0	40.6	—	—	—	—	—	1.6	10.9	18.6
393	—	—	—	69.0	3,017.0	2,476.0	65.5	—	—	—	—	—	1.0	4.1	14.8
394	—	—	—	75.0	3,283.0	2,695.0	71.2	—	—	—	—	—	1.1	4.5	16.1
395	—	—	—	75.0	3,597.0	2,781.0	72.0	—	—	—	—	—	0.7	8.4	10.5
396	—	—	—	82.0	3,914.0	3,026.0	78.3	—	—	—	—	—	0.7	9.2	11.4
397	—	—	—	61.0	3,480.0	2,633.0	61.1	—	—	—	—	—	0.9	9.6	19.0
398	—	—	—	67.0	3,799.0	2,875.0	66.7	—	—	—	—	—	1.0	10.5	20.8
399	—	—	—	67.0	2,946.0	2,582.0	56.8	—	—	—	—	—	0.9	9.1	25.3
400	—	—	—	72.0	3,163.0	2,772.0	61.0	—	—	—	—	—	1.0	9.8	27.1
401	—	—	—	—	—	—	61.1	—	—	—	—	—	0.3	11.4	17.8
402	—	—	—	—	—	—	65.6	—	—	—	—	—	0.3	12.2	19.1
403	—	—	—	67.0	2,946.0	2,531.0	65.2	—	—	—	—	—	1.0	5.0	15.8
404	—	—	—	72.0	3,163.0	2,717.0	70.0	—	—	—	—	—	1.1	5.4	17.0
405	—	—	—	65.0	3,237.0	2,369.0	59.0	—	—	—	—	—	0.8	6.9	21.9
406	—	—	—	70.0	3,489.0	2,554.0	63.6	—	—	—	—	—	0.9	7.4	23.6
407	—	—	—	68.0	3,017.0	2,656.0	62.6	—	—	—	—	—	0.7	4.6	23.2
408	—	—	—	75.0	3,307.0	2,912.0	68.2	—	—	—	—	—	0.8	5.1	25.4
409	—	—	—	68.0	3,006.0	2,776.0	16.6	—	—	—	—	—	12.1	9.3	6.2
410	—	—	—	75.0	3,297.0	3,044.0	18.2	—	—	—	—	—	13.2	10.2	6.8
411	47.0	—	—	73.0	3,381.0	2,761.0	34.3	23.0	—	—	6.0	15.0	8.8	5.4	5.7
412	51.0	—	—	81.0	3,727.0	3,044.0	37.9	25.0	—	—	7.0	17.0	9.6	6.0	6.3
413	—	—	—	65.0	2,883.0	2,523.0	34.6	23.0	—	—	5.0	17.0	9.1	1.4	5.8
414	—	—	—	72.0	3,196.0	2,797.0	38.3	25.0	—	—	6.0	19.0	10.1	1.5	6.5

TABLE 1   Composition of Important Feeds: Energy Values, Proximate Analyses, Plant Cell Wall Constituents, and Acid Detergent Fiber—*Continued*

Entry Number	Feed Name Description	International Feed Number	Dry Matter (%)	Ruminants TDN (%)	DE (Mcal/kg)	ME (Mcal/kg)	NE$_m$ (Mcal/kg)	NE$_g$ (Mcal/kg)	Dairy Cattle NE$_l$ (Mcal/kg)	Chickens ME$_n$ (kcal/kg)	TME (kcal/kg)	NE$_p$ (kcal/kg)
	GALLETA *Hilaria jamesii*											
415	fresh, stem-cured	2-05-594	71.0	34.0	1.49	1.19	0.73	0.14	0.74	—	—	—
416			100.0	48.0	2.12	1.69	1.03	0.19	1.06	—	—	—
	GELATIN											
417	process residue (Gelatin by-	5-14-503	90.0	72.0	3.17	2.80	1.69	1.12	1.65	2,138.0	—	—
418	products)		100.0	80.0	3.53	3.11	1.88	1.24	1.84	2,379.0	—	—
	GRAMA *Bouteloua* spp											
419	fresh, early vegetative	2-02-163	41.0	25.0	1.09	0.91	0.54	0.27	0.55	—	—	—
420			100.0	60.0	2.65	2.22	1.31	0.65	1.35	—	—	'—
421	fresh, mature	2-02-166	63.0	35.0	1.54	1.27	0.75	0.30	0.78	—	—	—
422			100.0	55.0	2.43	2.00	1.19	0.47	1.23	—	—	—
	GRAPE *Vitis* spp											
423	marc, dehy (Pomace)	1-02-208	91.0	30.0	1.32	0.93	0.73	—	0.63	1,558.0	—	—
424			100.0	33.0	1.46	1.02	0.81	—	0.69	1,715.0	—	—
	GROUNDNUT—SEE PEANUT											
	HEMICELLULOSE EXTRACT (MASONEX)											
425		4-08-030	76.0	46.0	2.02	1.70	1.00	0.50	1.03	1,506.0	—	1,214.0
426			100.0	60.0	2.65	2.22	1.31	0.65	1.35	1,973.0	—	1,591.0
	HOG MILLET—SEE MILLET, PROSO											
	HOMINY FEED—SEE CORN											
	JOHNSONGRASS—SEE SORGHUM, JOHNSONGRASS											
	KENTUCKY BLUEGRASS— SEE BLUEGRASS, KENTUCKY											
	LESPEDEZA, COMMON *Lespedeza striata*											
427	hay, sun-cured, midbloom	1-02-554	92.0	46.0	2.02	1.63	0.98	0.25	1.01	—	—	—
428			100.0	50.0	2.21	1.78	1.07	0.28	1.11	—	—	—
429	hay, sun-cured, full bloom	1-20-887	89.0	42.0	1.85	1.47	0.90	0.13	0.92	—	—	—
430			100.0	47.0	2.07	1.64	1.01	0.15	1.03	—	—	—
	LESPEDEZA, COMMON– LESPEDEZA, KOREAN *Lespedeza striata–Lespedeza stipulacea*											
431	fresh, late vegetative	2-26-028	32.0	19.0	0.83	0.70	0.41	0.20	0.42	—	—	—
432			100.0	59.0	2.60	2.18	1.28	0.62	1.33	—	—	—
433	fresh, early bloom	2-20-885	28.0	15.0	0.68	0.56	0.33	0.13	0.34	—	—	—
434			100.0	55.0	2.43	2.00	1.19	0.47	1.23	—	—	—
435	hay, sun-cured, late vege-	1-26-024	92.0	54.0	2.39	2.00	1.18	0.57	1.22	—	—	—
436	tative		100.0	59.0	2.60	2.18	1.28	0.62	1.33	—	—	—

	Horses			Swine				Plant Cell Wall Constituents							
Entry Num-ber	TDN (%)	DE (Mcal/ kg)	ME (Mcal/ kg)	TDN (%)	DE (kcal/ kg)	ME (kcal/ kg)	Crude Pro-tein (%)	Cell Walls (%)	Cell-ulose (%)	Hemi-cell-ulose (%)	Lig-nin (%)	Acid Deter-gent Fiber (%)	Crude Fiber (%)	Ether Ex-tract (%)	Ash (%)
415	—	—	—	—	—	—	3.9	—	—	—	—	—	23.3	1.3	11.4
416	—	—	—	—	—	—	5.5	—	—	—	—	—	33.0	1.8	16.2
417	—	—	—	—	2,801.0	2,138.0	87.6	—	—	—	—	—	—	0.0	—
418	—	—	—	—	3,177.0	2,379.0	97.4	—	—	—	—	—	—	0.1	—
419	—	—	—	—	—	—	5.4	—	—	—	—	—	11.2	0.8	4.6
420	—	—	—	—	—	—	13.1	—	—	—	—	—	27.2	2.0	11.3
421	—	—	—	—	—	—	4.1	—	—	—	—	—	20.7	1.1	7.2
422	—	—	—	—	—	—	6.5	—	—	—	—	—	32.7	1.7	11.4
423	—	—	—	—	—	—	11.8	50.0	—	—	32.0	49.0	29.0	7.2	9.3
424	—	—	—	—	—	—	13.0	55.0	—	—	35.0	54.0	31.9	7.9	10.3
425	—	—	—	—	—	—	0.6	—	—	—	—	—	0.8	0.3	3.1
426	—	—	—	—	—	—	0.7	—	—	—	—	—	1.0	0.4	4.1
427	43.0	1.74	1.43	—	—	—	11.5	—	—	—	—	—	26.4	2.3	4.5
428	47.0	1.90	1.56	—	—	—	12.6	—	—	—	—	—	28.8	2.5	4.9
429	43.0	1.73	1.42	—	—	—	12.8	—	—	—	—	—	27.4	1.9	5.0
430	48.0	1.93	1.58	—	—	—	14.3	—	—	—	—	—	30.7	2.1	5.6
431	—	—	—	—	—	—	5.2	—	—	—	—	—	10.2	—	—
432	—	—	—	—	—	—	16.4	—	—	—	—	—	32.0	—	—
433	—	—	—	—	—	—	4.6	—	—	—	—	—	9.0	—	—
434	—	—	—	—	—	—	16.4	—	—	—	—	—	32.0	—	—
435	—	—	—	—	—	—	16.4	—	—	—	—	—	22.1	—	—
436	—	—	—	—	—	—	17.8	—	—	—	—	—	24.0	—	—

TABLE 1　Composition of Important Feeds: Energy Values, Proximate Analyses, Plant Cell Wall Constituents, and Acid Detergent Fiber—*Continued*

Entry Number	Feed Name Description	International Feed Number	Dry Matter (%)	Ruminants TDN (%)	DE (Mcal/kg)	ME (Mcal/kg)	NE_m (Mcal/kg)	NE_g (Mcal/kg)	Dairy Cattle NE_l (Mcal/kg)	Chickens ME_n (kcal/kg)	TME (kcal/kg)	NE_p (kcal/kg)
437	hay, sun-cured, early bloom	1-26-025	93.0	51.0	2.26	1.86	1.10	0.44	1.14	—	—	—
438			100.0	55.0	2.43	2.00	1.19	0.47	1.23	—	—	—
439	hay, sun-cured, midbloom	1-26-026	93.0	47.0	2.05	1.65	1.00	0.26	1.03	—	—	—
440			100.0	50.0	2.21	1.78	1.07	0.28	1.11	—	—	—
441	hay, sun-cured, full bloom	1-26-027	93.0	44.0	1.93	1.53	0.94	0.14	0.96	—	—	—
442			100.0	47.0	2.07	1.64	1.01	0.15	1.03	—	—	—
	LESPEDEZA, KOREAN *Lespedeza stipulacea*											
443	fresh	2-02-598	30.0	19.0	0.85	0.72	0.43	0.24	0.44	—	—	—
444			100.0	64.0	2.82	2.40	1.41	0.78	1.45	—	—	—
445	hay, sun-cured	1-02-592	91.0	53.0	2.33	1.94	1.15	0.53	1.18	—	—	—
446			100.0	58.0	2.56	2.13	1.26	0.58	1.30	—	—	—
	LESPEDEZA, CHINESE *Lespedeza cuneata*											
447	fresh	2-02-611	35.0	19.0	0.84	0.69	0.41	0.16	0.43	—	—	—
448			100.0	55.0	2.43	2.00	1.19	0.47	1.23	—	—	—
449	hay, sun-cured, late vege-	1-09-172	92.0	41.0	1.82	1.42	0.89	0.05	0.90	—	—	—
450	tative		100.0	45.0	1.98	1.55	0.98	0.06	0.98	—	—	—
451	hay, sun-cured, early bloom	1-02-600	95.0	40.0	1.75	1.34	0.87	—	0.86	—	—	—
452			100.0	42.0	1.85	1.42	0.92	—	0.91	—	—	—
453	silage	3-02-614	30.0	14.0	0.60	0.47	0.30	0.02	0.30	—	—	—
454			100.0	45.0	1.98	1.55	0.98	0.06	0.98	—	—	—
	LIGNIN SULFONATE, CALCIUM											
455	dehy	8-16-028	97.0	8.0	0.34	—	0.88	—	0.08	—	—	—
456			100.0	8.0	0.35	—	0.91	—	0.08	—	—	—
	LINSEED—SEE FLAX											
	LIVERS											
457	meal	5-00-389	92.0	89.0	3.96	3.58	2.22	1.54	2.09	2,875.0	—	—
458			100.0	97.0	4.28	3.87	2.40	1.66	2.26	3,109.0	—	—
	MAIZE—SEE CORN											
	MANGELS—SEE BEET											
	MANURE—SEE CATTLE, SEE POULTRY											
	MASONEX—SEE HEMI-CELLULOSE EXTRACT											
	MEADOW PLANTS, INTERMOUNTAIN											
459	hay, sun-cured	1-03-181	95.0	55.0	2.43	2.03	1.20	0.55	1.24	—	—	—
460			100.0	58.0	2.56	2.13	1.26	0.58	1.30	—	—	—
	MEAT											
461	meal rendered	5-00-385	94.0	67.0	2.93	2.54	1.51	0.94	1.52	2,091.0	2,905.0	1,644.0
462			100.0	71.0	3.13	2.71	1.61	1.00	1.62	2,231.0	3,100.0	1,755.0

	Horses			Swine				Plant Cell Wall Constituents							
Entry Num- ber	TDN (%)	DE (Mcal/ kg)	ME (Mcal/ kg)	TDN (%)	DE (kcal/ kg)	ME (kcal/ kg)	Crude Pro- tein (%)	Cell Walls (%)	Cell- ulose (%)	Hemi- cell- ulose (%)	Lig- nin (%)	Acid Deter- gent Fiber (%)	Crude Fiber (%)	Ether Ex- tract (%)	Ash (%)
437	—	—	—	—	—	—	14.4	—	—	—	—	—	26.0	—	—
438	—	—	—	—	—	—	15.5	—	—	—	—	—	28.0	—	—
439	—	—	—	—	—	—	13.5	—	—	—	—	—	27.9	—	—
440	—	—	—	—	—	—	14.5	—	—	—	—	—	30.0	—	—
441	—	—	—	—	—	—	12.5	—	—	—	—	—	29.8	—	—
442	—	—	—	—	—	—	13.4	—	—	—	—	—	32.0	—	—
443	—	—	—	—	—	—	5.5	—	—	—	—	—	8.5	1.1	2.8
444	—	—	—	—	—	—	18.2	—	—	—	—	—	28.3	3.8	9.4
445	42.0	1.69	1.38	—	—	—	12.7	—	—	—	—	—	28.6	3.4	5.7
446	46.0	1.86	1.52	—	—	—	14.0	—	—	—	—	—	31.4	3.8	6.2
447	—	—	—	—	—	—	6.2	—	—	—	—	—	7.9	1.3	2.1
448	—	—	—	—	—	—	18.0	—	—	—	—	—	22.7	3.8	6.2
449	20.0	0.89	0.73	—	—	—	17.0	—	—	—	15.0	—	20.3	5.3	5.0
450	22.0	0.97	0.80	—	—	—	18.6	—	—	—	16.0	—	22.2	5.8	5.5
451	28.0	1.20	0.99	—	—	—	16.3	—	—	—	—	—	22.3	5.1	5.1
452	30.0	1.27	1.04	—	—	—	17.2	—	—	—	—	—	23.6	5.4	5.4
453	—	—	—	—	—	—	4.3	—	—	—	—	—	9.5	0.9	1.7
454	—	—	—	—	—	—	14.0	—	—	—	—	—	31.3	3.1	5.5
455	—	—	—	—	—	—	0.5	—	—	—	74.0	—	1.0	0.5	3.9
456	—	—	—	—	—	—	0.5	—	—	—	76.0	—	1.0	0.5	4.0
457	—	—	—	—	—	—	66.0	—	—	—	—	—	1.4	15.7	6.3
458	—	—	—	—	—	—	71.4	—	—	—	—	—	1.5	17.0	6.8
459	43.0	1.75	1.43	—	—	—	8.3	—	—	—	—	—	30.7	2.4	8.0
460	46.0	1.84	1.51	—	—	—	8.7	—	—	—	—	—	32.3	2.5	8.5
461	—	—	—	64.0	2,062.0	2,222.0	51.4	—	—	—	—	—	2.7	9.1	27.0
462	—	—	—	68.0	2,200.0	2,371.0	54.8	—	—	—	—	—	2.8	9.7	28.8

TABLE 1   Composition of Important Feeds: Energy Values, Proximate Analyses, Plant Cell Wall Constituents, and Acid Detergent Fiber—*Continued*

Entry Number	Feed Name Description	International Feed Number	Dry Matter (%)	Ruminants TDN (%)	DE (Mcal/kg)	ME (Mcal/kg)	NE$_m$ (Mcal/kg)	NE$_g$ (Mcal/kg)	Dairy Cattle NE$_l$ (Mcal/kg)	Chickens ME$_n$ (kcal/kg)	TME (kcal/kg)	NE$_p$ (kcal/kg)
463	with blood, meal rendered	5-00-386	92.0	67.0	2.92	2.54	1.51	0.95	1.51	2,672.0	2,981.0	1,781.0
464	(Tankage)		100.0	72.0	3.17	2.76	1.64	1.03	1.64	2,901.0	3,238.0	1,934.0
465	with blood with bone, meal	5-00-387	93.0	63.0	2.79	2.40	1.42	0.84	1.44	1,791.0	—	1,514.0
466	rendered (Tankage)		100.0	68.0	3.00	2.58	1.52	0.91	1.55	1,928.0	—	1,629.0
467	with bone, meal rendered	5-00-388	93.0	66.0	2.92	2.53	1.50	0.93	1.51	2,082.0	—	1,706.0
468			100.0	71.0	3.13	2.71	1.61	1.00	1.62	2,236.0	—	1,832.0
	MILK											
469	dehy (Cattle)	5-01-167	96.0	114.0	5.03	4.64	2.95	2.08	2.68	—	—	—
470			100.0	119.0	5.25	4.85	3.08	2.17	2.80	—	—	—
471	fresh (Cattle)	5-01-168	12.0	16.0	0.70	0.65	0.42	0.30	0.38	—	—	—
472			100.0	129.0	5.69	5.29	3.37	2.41	3.04	—	—	—
473	skimmed dehy (Cattle)	5-01-175	94.0	79.0	3.52	3.13	1.91	1.29	1.84	2,533.0	2,129.0	1,653.0
474			100.0	85.0	3.75	3.34	2.03	1.37	1.96	2,696.0	2,267.0	1,760.0
475	skimmed fresh (Cattle)	5-01-170	10.0	9.0	0.39	0.35	0.22	0.15	0.20	—	—	—
476			100.0	92.0	4.06	3.65	2.25	1.54	2.13	—	—	—
	MILLET, FOXTAIL *Setaria italica*											
477	fresh	2-03-101	28.0	18.0	0.78	0.67	0.39	0.21	0.40	—	—	—
478			100.0	63.0	2.78	2.36	1.39	0.75	1.42	—	—	—
479	grain	4-03-102	89.0	76.0	3.34	2.98	1.81	1.22	1.75	—	—	—
480			100.0	85.0	3.75	3.34	2.03	1.37	1.96	—	—	—
481	hay, sun-cured	1-03-099	87.0	51.0	2.27	1.90	1.12	0.54	1.16	—	—	—
482			100.0	59.0	2.60	2.18	1.28	0.62	1.33	—	—	—
	MILLET, PEARL—SEE PEARLMILLET											
	MILLET, PROSO *Panicum miliaceum*											
483	grain	4-03-120	90.0	75.0	3.33	2.96	1.80	1.21	1.74	2,898.0	—	—
484			100.0	84.0	3.70	3.29	2.00	1.35	1.94	3,222.0	—	—
	MOLASSES AND SYRUP											
485	beet, sugar, molasses, more	4-00-668	78.0	61.0	2.71	2.38	1.44	0.95	1.41	1,925.0	—	1,568.0
486	than 48% invert sugar more than 79.5 degrees brix		100.0	79.0	3.48	3.07	1.85	1.22	1.82	2,477.0	—	2,018.0
487	citrus, syrup (Citrus molasses)	4-01-241	68.0	51.0	2.24	1.96	1.17	0.75	1.16	—	—	—
488			100.0	75.0	3.31	2.89	1.73	1.11	1.72	—	—	—
489	sugarcane, molasses, dehy	4-04-695	94.0	66.0	2.91	2.52	1.49	0.91	1.51	2,706.0	—	—
490			100.0	70.0	3.09	2.67	1.58	0.97	1.60	2,866.0	—	—
491	sugarcane, molasses, more	4-04-696	75.0	54.0	2.37	2.05	1.22	0.77	1.23	1,927.0	—	1,563.0
492	than 46% invert sugar more than 79.5 degrees brix (Black strap)		100.0	72.0	3.17	2.76	1.64	1.03	1.64	2,585.0	—	2,098.0
	NAPIERGRASS *Pennisetum purpureum*											
493	fresh, late vegetative	2-03-158	20.0	11.0	0.49	0.41	0.24	0.10	0.25	—	—	—
494			100.0	55.0	2.43	2.00	1.19	0.47	1.23	—	—	—
495	fresh, late bloom	2-03-162	23.0	12.0	0.54	0.44	0.26	0.09	0.27	—	—	—
496			100.0	53.0	2.34	1.91	1.14	0.40	1.18	—	—	—

	Horses			Swine				Plant Cell Wall Constituents							
Entry Num- ber	TDN (%)	DE (Mcal/ kg)	ME (Mcal/ kg)	TDN (%)	DE (kcal/ kg)	ME (kcal/ kg)	Crude Pro- tein (%)	Cell Walls (%)	Cell- ulose (%)	Hemi- cell- ulose (%)	Lig- nin (%)	Acid Deter- gent Fiber (%)	Crude Fiber (%)	Ether Ex- tract (%)	Ash (%)
463	—	—	—	67.0	2,450.0	2,095.0	59.4	—	—	—	—	—	2.0	8.9	21.5
464	—	—	—	73.0	2,660.0	2,275.0	64.5	—	—	—	—	—	2.2	9.7	23.4
465	—	—	—	68.0	2,992.0	2,644.0	46.6	—	—	—	—	—	2.2	12.8	28.2
466	—	—	—	73.0	3,220.0	2,846.0	50.2	—	—	—	—	—	2.4	13.7	30.4
467	—	—	—	68.0	2,271.0	2,184.0	50.4	—	—	—	—	—	2.2	9.7	29.3
468	—	—	—	73.0	2,440.0	2,346.0	54.1	—	—	—	—	—	2.4	10.4	31.5
469	—	—	—	—	—	—	25.4	—	—	—	—	—	0.2	26.6	5.4
470	—	—	—	—	—	—	26.5	—	—	—	—	—	0.2	27.8	5.7
471	—	—	—	15.0	680.0	616.0	3.3	—	—	—	—	—	—	3.6	0.8
472	—	—	—	125.0	5,512.0	4,994.0	26.7	—	—	—	—	—	—	29.5	6.3
473	—	—	—	86.0	3,866.0	3,548.0	33.7	—	—	—	—	—	0.2	0.8	7.9
474	—	—	—	92.0	4,116.0	3,777.0	35.8	—	—	—	—	—	0.2	0.9	8.4
475	—	—	—	9.0	415.0	372.0	3.0	—	—	—	—	—	—	0.1	0.7
476	—	—	—	96.0	4,340.0	3,893.0	31.2	—	—	—	—	—	—	1.0	6.9
477	—	—	—	—	—	—	2.7	—	—	—	—	—	8.9	0.9	2.5
478	—	—	—	—	—	—	9.5	—	—	—	—	—	31.6	3.1	8.7
479	—	—	—	70.0	3,101.0	2,892.0	12.1	—	—	—	—	—	8.3	4.1	3.6
480	—	—	—	79.0	3,475.0	3,241.0	13.5	—	—	—	—	—	9.3	4.6	4.0
481	42.0	1.70	1.39	—	—	—	7.5	—	—	—	—	—	25.8	2.6	7.5
482	48.0	1.94	1.59	—	—	—	8.6	—	—	—	—	—	29.6	2.9	8.6
483	—	—	—	74.0	3,273.0	3,057.0	11.6	—	—	—	3.0	15.0	6.1	3.5	2.6
484	—	—	—	83.0	3,639.0	3,399.0	12.9	—	—	—	4.0	17.0	6.8	3.9	2.9
485	—	—	—	57.0	2,513.0	2,333.0	6.6	—	—	—	—	—	—	0.2	8.8
486	—	—	—	73.0	3,233.0	3,002.0	8.5	—	—	—	—	—	—	0.2	11.3
487	—	—	—	54.0	2,379.0	2,262.0	5.5	—	—	—	—	—	—	0.2	5.3
488	—	—	—	80.0	3,517.0	3,344.0	8.2	—	—	—	—	—	—	0.3	7.9
489	—	—	—	70.0	3,079.0	2,485.0	9.7	—	—	—	—	—	6.3	0.9	12.5
490	—	—	—	74.0	3,261.0	2,632.0	10.3	—	—	—	—	—	6.7	0.9	13.3
491	—	—	—	56.0	2,507.0	2,199.0	4.4	—	—	—	—	—	—	0.1	9.8
492	—	—	—	76.0	3,364.0	2,951.0	5.8	—	—	—	—	—	—	0.1	13.1
493	—	—	—	—	—	—	1.8	14.0	7.0	—	2.0	9.0	6.7	0.6	1.7
494	—	—	—	—	—	—	8.7	70.0	33.0	—	10.0	45.0	33.0	3.0	8.6
495	—	—	—	—	—	—	1.8	17.0	8.0	—	3.0	11.0	9.0	0.3	1.2
496	—	—	—	—	—	—	7.8	75.0	35.0	—	14.0	47.0	39.0	1.1	5.3

TABLE 1   Composition of Important Feeds: Energy Values, Proximate Analyses, Plant Cell Wall Constituents, and Acid Detergent Fiber—*Continued*

Entry Number	Feed Name Description	International Feed Number	Dry Matter (%)	Ruminants TDN (%)	DE (Mcal/ kg)	ME (Mcal/ kg)	NE$_m$ (Mcal/ kg)	NE$_g$ (Mcal/ kg)	Dairy Cattle NE$_l$ (Mcal/ kg)	Chickens ME$_n$ (kcal/ kg)	TME (kcal/ kg)	NE$_p$ (kcal/ kg)
	NEEDLEANDTHREAD *Stipa comata*											
497	fresh, stem-cured	2-07-989	92.0	45.0	1.99	1.59	0.97	0.22	0.99	—	—	—
498			100.0	49.0	2.16	1.73	1.05	0.23	1.08	—	—	—
	OATS *Avena sativa*											
499	breakfast cereal by-product,	4-03-303	91.0	86.0	3.80	3.43	2.12	1.46	2.00	3,158.0	3,800.0	2,536.0
500	less than 4% fiber (Feeding oat meal) (Oat middlings)		100.0	95.0	4.19	3.78	2.34	1.61	2.21	3,483.0	4,191.0	2,796.0
501	grain	4-03-309	89.0	68.0	3.02	2.65	1.59	1.04	1.57	2,543.0	3,264.0	1,735.0
502			100.0	77.0	3.40	2.98	1.79	1.17	1.77	2,862.0	3,674.0	1,953.0
503	grain, light less than 34.7 kg/	4-03-318	91.0	59.0	2.64	2.26	1.33	0.77	1.36	—	—	—
504	hl (Less than 27 lb/bushel)		100.0	66.0	2.91	2.49	1.47	0.85	1.50	—	—	—
505	grain, Pacific Coast	4-07-999	91.0	71.0	3.13	2.75	1.65	1.08	1.63	2,645.0	3,469.0	1,767.0
506			100.0	78.0	3.44	3.02	1.82	1.19	1.79	2,909.0	3,816.0	1,944.0
507	groats	4-03-331	90.0	85.0	3.71	3.35	2.07	1.42	1.96	3,251.0	—	2,523.0
508			100.0	94.0	4.14	3.74	2.31	1.59	2.18	3,630.0	—	2,817.0
509	hay, sun-cured	1-03-280	91.0	56.0	2.46	2.07	1.22	0.63	1.26	—	—	—
510			100.0	61.0	2.69	2.27	1.33	0.69	1.38	—	—	—
511	hulls	1-03-281	92.0	32.0	1.43	1.03	0.77	—	0.68	361.0	—	—
512			100.0	35.0	1.54	1.11	0.83	—	0.74	391.0	—	—
513	silage, late vegetative	3-20-898	23.0	15.0	0.66	0.56	0.33	0.19	0.34	—	—	—
514			100.0	65.0	2.87	2.45	1.44	0.82	1.47	—	—	—
515	silage, full bloom	3-07-893	58.0	34.0	1.53	1.28	0.76	0.38	0.78	—	—	—
516			100.0	60.0	2.65	2.22	1.31	0.65	1.35	—	—	—
517	silage, dough stage	3-03-296	35.0	20.0	0.88	0.73	0.43	0.19	0.45	—	—	—
518			100.0	57.0	2.51	2.09	1.23	0.55	1.28	—	—	—
519	straw	1-03-283	92.0	46.0	2.03	1.64	0.99	0.25	1.02	—	—	—
520			100.0	50.0	2.21	1.78	1.07	0.28	1.11	—	—	—
	ORANGE *Citrus sinensis*											
521	pulp without fines, dehy	4-01-254	88.0	72.0	3.19	2.82	1.71	1.14	1.67	—	—	—
522	(Orange pulp, dried)		100.0	82.0	3.62	3.20	1.94	1.30	1.89	—	—	—
	ORCHARDGRASS *Dactylis glomerata*											
523	fresh, early vegetative	2-03-439	23.0	17.0	0.74	0.64	0.38	0.24	0.38	—	—	—
524			100.0	72.0	3.17	2.76	1.64	1.03	1.64	—	—	—
525	fresh, early bloom	2-03-442	25.0	16.0	0.72	0.62	0.36	0.21	0.37	—	—	—
526			100.0	66.0	2.91	2.49	1.47	0.85	1.50	—	—	—
527	fresh, midbloom	2-03-443	31.0	17.0	0.77	0.64	0.38	0.17	0.39	—	—	—
528			100.0	57.0	2.51	2.09	1.23	0.55	1.28	—	—	—
529	fresh, milk stage	2-03-446	35.0	19.0	0.82	0.67	0.40	0.14	0.41	—	—	—
530			100.0	53.0	2.34	1.91	1.14	0.40	1.18	—	—	—
531	hay, sun-cured, early bloom	1-03-425	89.0	58.0	2.55	2.18	1.28	0.73	1.31	—	—	—
532			100.0	65.0	2.87	2.45	1.44	0.82	1.47	—	—	—
533	hay, sun-cured, late bloom	1-03-428	91.0	49.0	2.16	1.77	1.05	0.39	1.09	—	—	—
534			100.0	54.0	2.38	1.96	1.16	0.43	1.20	—	—	—
	PANGOLAGRASS *Digitaria decumbens*											
535	fresh	2-03-493	21.0	12.0	0.51	0.42	0.25	0.10	0.26	—	—	—
536			100.0	55.0	2.43	2.00	1.19	0.47	1.23	—	—	—

Entry Number	Horses TDN (%)	DE (Mcal/kg)	ME (Mcal/kg)	Swine TDN (%)	DE (kcal/kg)	ME (kcal/kg)	Crude Protein (%)	Cell Walls (%)	Cellulose (%)	Hemicellulose (%)	Lignin (%)	Acid Detergent Fiber (%)	Crude Fiber (%)	Ether Extract (%)	Ash (%)
497	—	—	—	—	—	—	3.7	76.0	33.0	37.0	6.0	40.0	—	5.0	19.4
498	—	—	—	—	—	—	4.1	83.0	36.0	40.0	6.0	43.0	—	5.4	21.1
499	—	—	—	79.0	3,480.0	3,427.0	14.8	—	—	—	—	—	3.5	6.4	2.3
500	—	—	—	87.0	3,838.0	3,779.0	16.4	—	—	—	—	—	3.9	7.0	2.5
501	66.0	—	—	64.0	2,825.0	2,676.0	11.8	28.0	10.0	13.0	2.0	14.0	10.8	4.8	3.1
502	74.0	—	—	72.0	3,180.0	3,012.0	13.3	32.0	11.0	15.0	3.0	16.0	12.1	5.4	3.4
503	—	—	—	61.0	2,696.0	2,517.0	11.9	—	—	—	—	—	14.4	4.5	4.2
504	—	—	—	67.0	2,973.0	2,776.0	13.1	—	—	—	—	—	15.9	4.9	4.6
505	—	—	—	69.0	3,030.0	2,623.0	9.1	—	—	—	—	—	11.2	5.0	3.8
506	—	—	—	76.0	3,333.0	2,886.0	10.0	—	—	—	—	—	12.3	5.5	4.2
507	—	—	—	84.0	3,718.0	2,928.0	15.8	—	—	—	—	—	2.5	6.2	2.1
508	—	—	—	94.0	4,152.0	3,269.0	17.7	—	—	—	—	—	2.8	6.9	2.4
509	43.0	1.73	1.42	—	—	—	8.5	60.0	—	24.0	5.0	33.0	27.8	2.4	7.0
510	47.0	1.89	1.55	—	—	—	9.3	66.0	—	26.0	6.0	36.0	30.4	2.6	7.6
511	23.0	0.99	0.81	27.0	1,198.0	870.0	3.6	72.0	28.0	—	7.0	39.0	30.9	1.6	6.1
512	25.0	1.07	0.87	29.0	1,295.0	940.0	3.9	78.0	30.0	—	8.0	42.0	33.4	1.8	6.6
513	—	—	—	—	—	—	2.9	13.0	—	—	—	—	6.9	0.6	1.5
514	—	—	—	—	—	—	12.8	58.0	—	—	—	—	29.9	2.5	6.5
515	—	—	—	—	—	—	5.6	—	—	—	—	—	19.5	1.9	4.8
516	—	—	—	—	—	—	9.6	—	—	—	—	—	33.7	3.2	8.3
517	—	—	—	—	—	—	3.5	—	—	—	—	—	11.6	1.4	2.4
518	—	—	—	—	—	—	10.0	—	—	—	—	—	33.0	4.1	6.9
519	44.0	1.77	1.45	—	—	—	4.1	64.0	37.0	—	13.0	43.0	37.3	2.1	7.2
520	48.0	1.92	1.58	—	—	—	4.4	70.0	40.0	—	14.0	47.0	40.5	2.2	7.8
521	—	—	—	72.0	3,154.0	2,974.0	7.5	19.0	—	—	—	—	8.4	1.7	3.7
522	—	—	—	81.0	3,578.0	3,374.0	8.5	21.0	—	—	—	—	9.6	1.9	4.2
523	—	—	—	—	—	—	4.3	13.0	6.0	6.0	1.0	7.0	5.8	1.1	2.6
524	—	—	—	—	—	—	18.4	55.0	25.0	24.0	3.0	31.0	24.7	4.9	11.3
525	—	—	—	—	—	—	4.0	15.0	7.0	6.0	1.0	8.0	7.4	0.9	1.8
526	—	—	—	—	—	—	16.0	60.0	28.0	25.0	4.0	33.0	30.0	3.7	7.2
527	—	—	—	—	—	—	3.4	21.0	10.0	8.0	2.0	13.0	10.2	1.1	2.3
528	—	—	—	—	—	—	11.0	68.0	33.0	27.0	6.0	41.0	33.5	3.5	7.5
529	—	—	—	—	—	—	2.9	25.0	13.0	9.0	3.0	15.0	12.3	1.3	2.1
530	—	—	—	—	—	—	8.4	71.0	38.0	27.0	8.0	44.0	35.2	3.7	6.0
531	42.0	1.70	1.40	—	—	—	13.4	54.0	26.0	24.0	4.0	30.0	27.6	2.5	7.8
532	48.0	1.91	1.57	—	—	—	15.0	61.0	29.0	27.0	5.0	34.0	31.0	2.8	8.7
533	44.0	1.77	1.45	—	—	—	7.6	65.0	35.0	24.0	8.0	41.0	33.6	3.1	9.2
534	49.0	1.96	1.61	—	—	—	8.4	72.0	39.0	27.0	9.0	45.0	37.1	3.4	10.1
535	—	—	—	11.0	499.0	469.0	2.1	—	—	—	1.0	8.0	6.4	0.5	2.0
536	—	—	—	54.0	2,389.0	2,244.0	10.3	—	—	—	5.0	38.0	30.5	2.3	9.6

TABLE 1   Composition of Important Feeds: Energy Values, Proximate Analyses, Plant Cell Wall Constituents, and Acid Detergent Fiber—*Continued*

Entry Number	Feed Name Description	International Feed Number	Dry Matter (%)	Ruminants TDN (%)	DE (Mcal/kg)	ME (Mcal/kg)	NE$_m$ (Mcal/kg)	NE$_g$ (Mcal/kg)	Dairy Cattle NE$_l$ (Mcal/kg)	Chickens ME$_n$ (kcal/kg)	TME (kcal/kg)	NE$_p$ (kcal/kg)
537	hay, sun-cured	1-09-459	88.0	43.0	1.90	1.53	0.93	0.21	0.95	—	—	—
538			100.0	49.0	2.16	1.73	1.05	0.23	1.08	—	—	—
539	hay, sun-cured, 15 to 28	1-10-638	91.0	46.0	2.05	1.66	1.00	0.29	1.03	—	—	—
540	days' growth		100.0	51.0	2.25	1.82	1.10	0.32	1.13	—	—	—
541	hay, sun-cured, 29 to 42	1-26-214	91.0	41.0	1.81	1.41	0.89	0.05	0.89	—	—	—
542	days' growth		100.0	45.0	1.98	1.55	0.98	0.06	0.98	—	—	—
543	hay, sun-cured, 43 to 56	1-29-573	91.0	36.0	1.61	1.21	0.81	—	0.78	—	—	—
544	days' growth		100.0	40.0	1.76	1.33	0.89	—	0.86	—	—	—
	PAPER											
545	corrugated	1-28-257	93.0	64.0	2.83	2.44	1.44	0.87	1.46	—	—	—
546			100.0	69.0	3.04	2.62	1.55	0.94	1.57	—	—	—
	PEA *Pisum* spp											
547	seeds	5-03-600	89.0	77.0	3.41	3.05	1.86	1.26	1.79	2,123.0	—	1,542.0
548			100.0	87.0	3.84	3.42	2.09	1.42	2.01	2,385.0	—	1,732.0
549	straw	1-03-577	87.0	40.0	1.76	1.39	0.87	0.09	0.88	—	—	—
550			100.0	46.0	2.03	1.60	0.99	0.10	1.01	—	—	—
551	vines without seeds, silage	3-03-596	25.0	14.0	0.62	0.51	0.30	0.13	0.31	—	—	—
552			100.0	57.0	2.51	2.09	1.23	0.55	1.28	—	—	—
	PEANUT *Arachis hypogaea*											
553	hay, sun-cured	1-03-619	91.0	50.0	2.20	1.82	1.08	0.43	1.12	—	—	—
554			100.0	55.0	2.43	2.00	1.19	0.47	1.23	—	—	—
555	hulls	1-08-028	91.0	20.0	0.89	0.48	0.70	—	0.38	—	—	—
556			100.0	22.0	0.97	0.53	0.76	—	0.42	—	—	—
557	kernels, meal mech extd	5-03-649	93.0	77.0	3.39	3.01	1.82	1.22	1.77	2,662.0	—	1,898.0
558	(Peanut meal)		100.0	83.0	3.66	3.25	1.97	1.32	1.91	2,874.0	—	2,049.0
559	kernels, meal solv extd	5-03-650	92.0	71.0	3.12	2.74	1.64	1.07	1.63	2,693.0	—	1,967.0
560	(Peanut meal)		100.0	77.0	3.40	2.98	1.79	1.17	1.77	2,928.0	—	2,138.0
	PEARLMILLET *Pennisetum glaucum*											
561	fresh	2-03-115	21.0	13.0	0.57	0.48	0.28	0.14	0.29	—	—	—
562			100.0	61.0	2.69	2.27	1.33	0.69	1.38	—	—	—
563	silage	3-20-903	30.0	18.0	0.77	0.65	0.38	0.18	0.40	—	—	—
564			100.0	59.0	2.60	2.18	1.28	0.62	1.33	—	—	—
	PINEAPPLE *Ananas comosus*											
565	aerial part without fruit,	1-13-309	89.0	54.0	2.39	2.01	1.18	0.61	1.22	—	—	—
566	sun-cured (Pineapple hay)		100.0	61.0	2.69	2.27	1.33	0.69	1.38	—	—	—
567	process residue, dehy	4-03-722	87.0	59.0	2.61	2.25	1.33	0.79	1.35	—	—	—
568	(Pineapple bran)		100.0	68.0	3.00	2.58	1.52	0.91	1.55	—	—	—
	POTATO *Solanum tuberosum*											
569	process residue, dehy	4-03-775	89.0	79.0	3.52	3.16	1.94	1.33	1.85	—	—	—
570			100.0	90.0	3.97	3.56	2.19	1.49	2.09	—	—	—
571	tubers, dehy	4-07-850	91.0	74.0	3.26	2.88	1.74	1.16	1.70	2,825.0	—	1,988.0
572			100.0	81.0	3.57	3.16	1.91	1.27	1.87	3,098.0	—	2,180.0
573	tubers, fresh	4-03-787	23.0	19.0	0.84	0.74	0.45	0.30	0.44	—	—	—
574			100.0	81.0	3.57	3.16	1.91	1.27	1.87	—	—	—
575	tubers, silage	4-03-768	25.0	20.0	0.89	0.79	0.48	0.32	0.47	—	—	—
576			100.0	82.0	3.62	3.20	1.94	1.30	1.89	—	—	—
577	tubers, boiled silage	4-03-767	23.0	18.0	0.77	0.68	0.40	0.26	0.40	—	—	—
578			100.0	75.0	3.31	2.89	1.73	1.11	1.72	—	—	—

	Horses			Swine				Plant Cell Wall Constituents							
Entry Num-ber	TDN (%)	DE (Mcal/ kg)	ME (Mcal/ kg)	TDN (%)	DE (kcal/ kg)	ME (kcal/ kg)	Crude Pro-tein (%)	Cell Walls (%)	Cell-ulose (%)	Hemi-cell-ulose (%)	Lig-nin (%)	Acid Deter-gent Fiber (%)	Crude Fiber (%)	Ether Ex-tract (%)	Ash (%)
537	38.0	1.54	1.26	—	—	—	6.7	—	—	—	—	—	27.4	1.5	11.7
538	43.0	1.75	1.43	—	—	—	7.6	—	—	—	—	—	31.3	1.7	13.3
539	—	—	—	—	—	—	10.5	64.0	30.0	—	5.0	37.0	30.9	2.0	7.7
540	—	—	—	—	—	—	11.5	70.0	33.0	—	6.0	41.0	34.0	2.2	8.5
541	—	—	—	—	—	—	6.5	66.0	32.0	—	6.0	39.0	32.8	1.8	7.3
542	—	—	—	—	—	—	7.1	73.0	35.0	—	6.0	43.0	36.0	2.0	8.0
543	—	—	—	—	—	—	5.0	70.0	34.0	—	6.0	42.0	34.6	1.8	6.9
544	—	—	—	—	—	—	5.5	77.0	37.0	—	7.0	46.0	38.0	2.0	7.6
545	—	—	—	—	—	—	—	—	—	—	—	—	—	—	—
546	—	—	—	—	—	—	—	—	—	—	—	—	—	—	—
547	69.0	—	—	76.0	3,346.0	3,041.0	22.5	—	—	—	—	—	6.1	1.2	3.0
548	78.0	—	—	85.0	3,759.0	3,416.0	25.3	—	—	—	—	—	6.9	1.4	3.3
549	36.0	1.47	1.21	—	—	—	7.8	—	—	—	—	—	34.3	1.5	5.7
550	42.0	1.69	1.39	—	—	—	8.9	—	—	—	—	—	39.5	1.8	6.5
551	—	—	—	—	—	—	3.2	14.0	8.0	—	2.0	12.0	7.3	0.8	2.2
552	—	—	—	—	—	—	13.1	59.0	34.0	—	9.0	49.0	29.8	3.3	9.0
553	44.0	1.77	1.45	—	—	—	9.8	—	—	—	—	—	30.2	3.1	7.8
554	49.0	1.95	1.60	—	—	—	10.8	—	—	—	—	—	33.2	3.4	8.6
555	—	—	—	—	—	—	7.1	67.0	36.0	—	21.0	59.0	57.3	1.8	3.8
556	—	—	—	—	—	—	7.8	74.0	40.0	—	23.0	65.0	62.9	2.0	4.2
557	—	—	—	80.0	4,107.0	3,466.0	48.1	13.0	4.0	—	6.0	6.9	5.8	5.1
558	—	—	—	86.0	4,434.0	3,741.0	52.0	14.0	5.0	—	6.0	7.5	6.3	5.5
559	—	—	—	79.0	3,496.0	3,031.0	48.1	—	—	—	—	—	9.9	1.3	5.8
560	—	—	—	86.0	3,800.0	3,295.0	52.3	—	—	—	—	—	10.8	1.4	6.3
561	—	—	—	—	—	—	1.8	—	—	—	—	—	6.6	0.5	2.1
562	—	—	—	—	—	—	8.5	—	—	—	—	—	31.5	2.2	10.0
563	—	—	—	—	—	—	2.7	—	—	—	—	—	11.3	—	—
564	—	—	—	—	—	—	9.2	—	—	—	—	—	38.0	—	—
565	42.0	1.69	1.39	—	—	—	6.9	—	—	—	—	—	26.3	2.5	5.5
566	47.0	1.91	1.56	—	—	—	7.8	—	—	—	—	—	29.6	2.8	6.1
567	—	—	—	60.0	2,667.0	2,536.0	4.0	64.0	—	—	6.0	32.0	18.2	1.3	3.0
568	—	—	—	69.0	3,063.0	2,913.0	4.6	73.0	—	—	7.0	37.0	20.9	1.5	3.5
569	—	—	—	76.0	3,367.0	3,175.0	7.4	—	—	—	—	—	6.5	0.3	3.0
570	—	—	—	86.0	3,791.0	3,575.0	8.4	—	—	—	—	—	7.3	0.4	3.4
571	—	—	—	75.0	3,301.0	3,265.0	8.1	—	—	—	—	—	2.1	0.5	7.9
572	—	—	—	82.0	3,620.0	3,581.0	8.9	—	—	—	—	—	2.3	0.5	8.7
573	—	—	—	20.0	876.0	825.0	2.2	—	—	—	—	—	0.6	0.1	1.1
574	—	—	—	85.0	3,737.0	3,516.0	9.5	—	—	—	—	—	2.4	0.4	4.8
575	—	—	—	22.0	963.0	910.0	1.9	—	—	—	—	—	1.0	0.1	1.4
576	—	—	—	89.0	3,902.0	3,686.0	7.6	—	—	—	—	—	4.0	0.4	5.5
577	—	—	—	21.0	917.0	865.0	1.9	—	—	—	—	—	0.7	0.1	1.5
578	—	—	—	89.0	3,915.0	3,694.0	8.2	—	—	—	—	—	3.2	0.4	6.5

TABLE 1   Composition of Important Feeds: Energy Values, Proximate Analyses, Plant Cell Wall Constituents, and Acid Detergent Fiber—*Continued*

Entry Number	Feed Name Description	International Feed Number	Dry Matter (%)	Ruminants					Dairy Cattle NE_l (Mcal/ kg)	Chickens		
				TDN (%)	DE (Mcal/ kg)	ME (Mcal/ kg)	NE_m (Mcal/ kg)	NE_g (Mcal/ kg)		ME_n (kcal/ kg)	TME (kcal/ kg)	NE_p (kcal/ kg)
579	vines, silage	3-03-765	15.0	8.0	0.37	0.31	0.18	0.08	0.19	—	—	—
580			100.0	57.0	2.51	2.09	1.23	0.55	1.28	—	—	—
	POULTRY											
581	by-product, meal rendered	5-03-798	93.0	74.0	3.26	2,87	1.73	1.14	1.70	2,851.0	4,171.0	2,016.0
582	(Viscera with feet with heads)		100.0	79.0	3.48	3.07	1.85	1.22	1.82	3,051.0	4,463.0	2,158.0
583	feathers, hydrolyzed	5-03-795	93.0	65.0	2.87	2.48	1.47	0.90	1.48	2,427.0	3,941.0	1,521.0
584			100.0	70.0	3.09	2.67	1.58	0.97	1.60	2,609.0	4,238.0	1,636.0
585	manure and litter	5-05-587	89.0	59.0	2.60	2.22	1.31	0.76	1.34	—	—	—
586			100.0	66.0	2.91	2.49	1.47	0.85	1.50	—	—	—
587	manure, dehy	5-14-015	90.0	52.0	2.31	1.93	1.14	0.53	1.17	1,031.0	—	—
588			100.0	58.0	2.56	2.13	1.26	0.58	1.30	1,142.0	—	—
	PRAIRIE PLANTS, MIDWEST											
589	hay, sun-cured	1-03-191	92.0	47.0	2.06	1.67	1.00	0.29	1.04	—	—	—
590			100.0	51.0	2.25	1.82	1.10	0.32	1.13	—	—	—
	PRICKLYPEAR *Opuntia* spp											
591	fresh	2-01-061	17.0	9.0	0.42	0.35	0.21	0.09	0.21	—	—	—
592			100.0	57.0	2.51	2.09	1.23	0.55	1.28	—	—	—
	PROPYLENE GLYCOL											
593		8-03-809	100.0	158.0	6.95	6.57	4.15	3.15	3.74	—	—	—
594			100.0	158.0	6.97	6.59	4.16	3.16	3.75	—	—	—
	RAPE *Brassica* spp											
595	fresh, early vegetative	2-03-865	18.0	15.0	0.65	0.58	0.35	0.23	0.34	—	—	—
596			100.0	81.0	3.57	3.16	1.91	1.27	1.87			
597	fresh, early bloom	2-03-866	11.0	8.0	0.37	0.33	0.20	0.13	0.19	—	—	—
598			100.0	75.0	3.31	2.89	1.73	1.11	1.72	—	—	—
599	seeds, meal mech extd	5-03-870	92.0	70.0	3.08	2.70	1.62	1.05	1.60	2,003.0	2,216.0	1,076.0
600			100.0	76.0	3.35	2.93	1.76	1.14	1.74	2,177.0	2,410.0	1,170.0
601	seeds, meal solv extd	5-03-871	91.0	63.0	2.77	2.39	1.41	0.86	1.43	1,751.0	2,103.0	1,088.0
602			100.0	69.0	3.04	2.62	1.55	0.94	1.57	1,924.0	2,310.0	1,196.0
	RAPE, SUMMER *Brassica napus*											
603	seeds, meal mech extd	5-08-136	94.0	70.0	3.07	2.68	1.60	1.02	1.59	—	—	—
604			100.0	74.0	3.26	2.85	1.70	1.08	1.69	—	—	—
605	seeds, meal prepressed	5-08-135	92.0	69.0	3.04	2.66	1.59	1.02	1.58	—	—	—
606	solv extd		100.0	75.0	3.31	2.89	1.73	1.11	1.72	—	—	—
	REDTOP *Agrostis alba*											
607	fresh	2-03-897	29.0	18.0	0.81	0.69	0.41	0.22	0.42	—	—	—
608			100.0	63.0	2.78	2.36	1.39	0.75	1.42	—	—	—
609	hay, sun-cured, midbloom	1-03-886	94.0	54.0	2.37	1.97	1.17	0.52	1.21	—	—	—
610			100.0	57.0	2.51	2.09	1.23	0.55	1.28	—	—	—
	RICE *Oryza sativa*											
611	bran with germ (Rice bran)	4-03-928	91.0	63.0	2.80	2.42	1.43	0.88	1.45	2,106.0	—	1,713.0
612			100.0	70.0	3.09	2.67	1.58	0.97	1.60	2,323.0	—	1,889.0
613	grain, ground (Ground rough	4-03-938	89.0	70.0	3.10	2.73	1.64	1.08	1.61	2,664.0	—	1,784.0
614	rice) (Ground paddy rice)		100.0	79.0	3.48	3.07	1.85	1.22	1.82	2,998.0	—	2,008.0

| | Horses | | | Swine | | | | Plant Cell Wall Constituents | | | | | | | |
Entry Num- ber	TDN (%)	DE (Mcal/ kg)	ME (Mcal/ kg)	TDN (%)	DE (kcal/ kg)	ME (kcal/ kg)	Crude Pro- tein (%)	Cell Walls (%)	Cell- ulose (%)	Hemi- cell- ulose (%)	Lig- nin (%)	Acid Deter- gent Fiber (%)	Crude Fiber (%)	Ether Ex- tract (%)	Ash (%)
579	—	—	—	—	—	—	2.3	—	—	—	—	—	3.4	0.5	2.8
580	—	—	—	—	—	—	15.6	—	—	—	—	—	23.0	3.7	19.1
581	—	—	—	75.0	3,088.0	2,858.0	58.7	—	—	—	—	—	2.3	13.1	15.7
582	—	—	—	81.0	3,305.0	3,058.0	62.8	—	—	—	—	—	2.4	14.1	16.8
583	—	—	—	62.0	2,731.0	2,215.0	84.9	—	—	—	—	—	1.4	2.9	3.5
584	—	—	—	67.0	2,936.0	2,382.0	91.3	—	—	—	—	—	1.5	3.2	3.8
585	—	—	—	—	—	—	21.9	—	—	—	—	—	14.4	2.7	19.7
586	—	—	—	—	—	—	24.5	—	—	—	—	—	16.1	3.0	22.0
587	—	—	—	—	—	—	25.5	34.0	—	—	2.0	14.0	11.9	2.2	27.2
588	—	—	—	—	—	—	28.2	38.0	—	—	2.0	15.0	13.2	2.4	30.1
589	40.0	1.63	1.33	—	—	—	5.3	—	—	—	—	—	31.1	2.2	6.5
590	44.0	1.78	1.46	—	—	—	5.8	—	—	—	—	—	34.0	2.4	7.1
591	—	—	—	10.0	432.0	411.0	0.8	5.0	—	—	1.0	4.0	2.3	0.3	3.4
592	—	—	—	58.0	2,578.0	2,450.0	4.8	30.0	—	—	8.0	23.0	13.5	1.9	20.1
593	—	—	—	—	—	—	—	—	—	—	—	—	—	—	—
594	—	—	—	—	—	—	—	—	—	—	—	—	—	—	—
595	—	—	—	—	—	—	3.0	—	—	—	—	—	2.4	0.7	2.1
596	—	—	—	—	—	—	16.4	—	—	—	—	—	13.0	4.0	11.4
597	—	—	—	—	—	—	2.7	—	—	—	—	—	1.8	0.4	1.6
598	—	—	—	—	—	—	23.5	—	—	—	—	—	15.8	3.8	14.0
599	—	—	—	67.0	2,933.0	2,612.0	35.6	—	—	—	—	—	12.0	7.3	6.9
600	—	—	—	72.0	3,189.0	2,840.0	38.7	—	—	—	—	—	13.1	7.9	7.5
601	—	—	—	65.0	2,878.0	2,672.0	37.0	—	—	—	—	—	12.0	1.7	6.8
602	—	—	—	72.0	3,161.0	2,935.0	40.6	—	—	—	—	—	13.2	1.8	7.5
603	—	—	—	—	—	—	35.2	—	—	—	—	—	15.5	7.0	6.8
604	—	—	—	—	—	—	37.4	—	—	—	—	—	16.5	7.4	7.2
605	—	—	—	—	—	—	40.5	—	—	—	—	—	9.3	1.1	7.2
606	—	—	—	—	—	—	44.0	—	—	—	—	—	10.1	1.2	7.8
607	—	—	—	—	—	—	3.4	19.0	—	6.0	2.0	—	7.8	1.2	2.4
608	—	—	—	—	—	—	11.6	64.0	—	19.0	8.0	—	26.7	3.9	8.1
609	45.0	1.80	1.48	—	—	—	11.0	—	—	—	—	—	29.0	2.5	6.1
610	47.0	1.91	1.56	—	—	—	11.7	—	—	—	—	—	30.7	2.6	6.5
611	—	—	—	72.0	3,384.0	3,070.0	12.7	30.0	10.0	14.0	—	16.0	11.6	13.7	11.6
612	—	—	—	79.0	3,733.0	3,387.0	14.1	33.0	11.0	15.0	—	18.0	12.8	15.1	12.8
613	—	—	—	62.0	2,755.0	2,722.0	7.9	—	—	—	—	—	8.9	1.7	4.7
614	—	—	—	70.0	3,100.0	3,063.0	8.9	—	—	—	—	—	10.0	1.9	5.3

TABLE 1   Composition of Important Feeds: Energy Values, Proximate Analyses, Plant Cell Wall Constituents, and Acid Detergent Fiber—*Continued*

Entry Number	Feed Name Description	International Feed Number	Dry Matter (%)	Ruminants TDN (%)	DE (Mcal/kg)	ME (Mcal/kg)	NE$_m$ (Mcal/kg)	NE$_g$ (Mcal/kg)	Dairy Cattle NE$_l$ (Mcal/kg)	Chickens ME$_n$ (kcal/kg)	TME (kcal/kg)	NE$_p$ (kcal/kg)
615	grain, polished and broken	4-03-932	89.0	79.0	3.47	3.11	1.91	1.30	1.82	3,092.0	—	2,500.0
616	(Brewers rice)		100.0	89.0	3.92	3.51	2.15	1.47	2.06	3,493.0	—	2,824.0
617	groats, polished (Rice,	4-03-942	89.0	78.0	3.44	3.07	1.88	1.28	1.80	3,087.0	—	—
618	polished)		100.0	88.0	3.88	3.47	2.12	1.45	2.04	3,483.0	—	—
619	hulls	1-08-075	92.0	11.0	0.49	0.08	0.78	—	0.16	79.0	1,208.0	113.0
620			100.0	12.0	0.53	0.08	0.84	—	0.17	86.0	1,310.0	122.0
621	polishings	4-03-943	90.0	81.0	3.58	3.21	1.97	1.35	1.88	3,030.0	—	2,090.0
622			100.0	90.0	3.97	3.56	2.19	1.49	2.09	3,357.0	—	2,315.0
	RUSSIANTHISTLE, TUMBLING *Salsola kali tenuifolia*											
623	fresh, stem-cured	2-08-000	88.0	44.0	1.94	1.56	0.95	0.24	0.97	—	—	—
624			100.0	50.0	2.12	1.78	1.07	0.28	1.11	—	—	—
625	hay, sun-cured	1-03-988	86.0	39.0	1.74	1.37	0.85	0.09	0.87	—	—	—
626			100.0	46.0	2.03	1.60	0.99	0.10	1.01	—	—	—
	RYE *Secale cereale*											
627	distillers grains, dehy	5-04-023	92.0	56.0	2.47	2.08	1.23	0.63	1.26	—	—	—
628			100.0	61.0	2.69	2.27	1.33	0.69	1.38	—	—	—
629	fresh	2-04-018	24.0	17.0	0.73	0.63	0.37	0.23	0.38	—	—	—
630			100.0	69.0	3.04	2.62	1.55	0.94	1.57	—	—	—
631	fresh, early vegetative	2-04-013	16.0	11.0	0.47	0.41	0.24	0.15	0.24	—	—	—
632			100.0	69.0	3.04	2.62	1.55	0.94	1.57	—	—	—
633	grain	4-04-047	88.0	73.0	3.24	2.88	1.75	1.18	1.70	2,626.0	3,185.0	2,074.0
634			100.0	84.0	3.70	3.29	2.00	1.35	1.94	3,001.0	3,640.0	2,370.0
635	flour by-product, less than	4-04-032	89.0	61.0	2.70	2.33	1.38	0.83	1.39	—	—	—
636	4.5% fiber (Rye middlings)		100.0	69.0	3.04	2.62	1.55	0.94	1.57	—	—	—
637	flour by-product, less than	4-04-031	89.0	73.0	3.23	2.86	1.73	1.16	1.69	—	—	—
638	8.5% fiber (Rye middlings)		100.0	82.0	3.62	3.20	1.94	1.30	1.89	—	—	—
639	mill run, less than 9.5%	4-04-034	90.0	68.0	2.98	2.61	1.56	1.00	1.55	—	—	—
640	fiber (Rye feed)		100.0	75.0	3.31	2.89	1.73	1.11	1.72	—	—	—
641	silage	3-04-020	32.0	17.0	0.75	0.61	0.37	0.13	0.38	—	—	—
642			100.0	53.0	2.34	1.91	1.14	0.40	1.18	—	—	—
643	straw	1-04-007	90.0	39.0	1.74	1.36	0.86	0.01	0.86	—	—	—
644			100.0	44.0	1.94	1.51	0.96	0.01	0.96	—	—	—
	RYEGRASS, ITALIAN *Lolium multiflorum*											
645	fresh	2-04-073	25.0	15.0	0.65	0.55	0.32	0.16	0.33	—	—	—
646			100.0	60.0	2.65	2.22	1.31	0.65	1.35	—	—	—
647	hay, sun-cured, early vege-	1-04-064	89.0	61.0	2.68	2.30	1.36	0.81	1.38	—	—	—
648	tative		100.0	68.0	3.00	2.58	1.52	0.91	1.55	—	—	—
649	hay, sun-cured, late vege-	1-04-065	86.0	53.0	2.34	1.98	1.16	0.62	1.20	—	—	—
650	tative		100.0	62.0	2.73	2.31	1.36	0.72	1.40	—	—	—
651	hay, sun-cured, early bloom	1-04-066	83.0	45.0	1.99	1.63	0.97	0.36	1.00	—	—	—
652			100.0	54.0	2.38	1.96	1.16	0.43	1.20	—	—	—
	RYEGRASS, PERENNIAL *Lolium perenne*											
653	fresh	2-04-086	27.0	18.0	0.80	0.69	0.41	0.24	0.41	—	—	—
654			100.0	68.0	3.00	2.58	1.52	0.91	1.55	—	—	—
655	hay, sun-cured	1-04-077	86.0	55.0	2.43	2.07	1.22	0.68	1.25	—	—	—
656			100.0	64.0	2.82	2.40	1.41	0.78	1.45	—	—	—

| | Horses | | | Swine | | | | Plant Cell Wall Constituents | | | | | | | |
Entry Number	TDN (%)	DE (Mcal/ kg)	ME (Mcal/ kg)	TDN (%)	DE (kcal/ kg)	ME (kcal/ kg)	Crude Protein (%)	Cell Walls (%)	Cellulose (%)	Hemicellulose (%)	Lignin (%)	Acid Detergent Fiber (%)	Crude Fiber (%)	Ether Extract (%)	Ash (%)
615	—	—	—	79.0	4,162.0	3,387.0	7.6	14.0	0.0	13.0	—	1.0	0.6	0.7	0.8
616	—	—	—	89.0	4,702.0	3,826.0	8.6	16.0	1.0	15.0	—	1.0	0.7	0.8	0.8
617	—	—	—	86.0	3,776.0	3,562.0	7.2	—	—	—	—	—	0.4	0.4	0.5
618	—	—	—	97.0	4,261.0	4,020.0	8.2	—	—	—	—	—	0.4	0.5	0.6
619	—	—	—	—	—	—	3.0	76.0	30.0	—	15.0	66.0	39.6	0.7	19.0
620	—	—	—	—	—	—	3.3	82.0	33.0	—	16.0	72.0	42.9	0.8	20.6
621	—	—	—	88.0	3,936.0	3,443.0	12.1	—	—	—	—	—	3.2	12.5	7.5
622	—	—	—	97.0	4,360.0	3,814.0	13.4	—	—	—	—	—	3.6	13.9	8.3
623	—	—	—	—	—	—	9.9	56.0	—	—	10.0	39.0	—	2.6	13.4
624	—	—	—	—	—	—	11.2	64.0	—	—	11.0	44.0	—	3.0	15.2
625	43.0	1.70	1.40	—	—	—	10.7	—	—	—	—	—	24.4	1.8	13.2
626	50.0	1.98	1.63	—	—	—	12.4	—	—	—	—	—	28.4	2.1	15.4
627	—	—	—	—	—	—	21.6	—	—	—	—	—	12.3	7.2	2.3
628	—	—	—	—	—	—	23.5	—	—	—	—	—	13.4	7.8	2.5
629	—	—	—	—	—	—	3.8	—	—	—	—	—	6.8	0.9	1.9
630	—	—	—	—	—	—	15.9	—	—	—	—	—	28.5	3.7	8.1
631	—	—	—	—	—	—	4.3	—	—	—	—	—	—	—	—
632	—	—	—	—	—	—	28.0	—	—	—	—	—	—	—	—
633	—	—	—	75.0	3,251.0	2,911.0	12.1	—	—	—	—	—	2.2	1.5	1.6
634	—	—	—	86.0	3,716.0	3,327.0	13.8	—	—	—	—	—	2.5	1.7	1.9
635	—	—	—	57.0	2,497.0	2,338.0	10.4	—	—	—	—	—	16.5	2.6	3.2
636	—	—	—	64.0	2,813.0	2,634.0	11.8	—	—	—	—	—	18.6	3.0	3.6
637	—	—	—	73.0	3,220.0	2,972.0	16.2	—	—	—	—	—	4.8	3.2	3.7
638	—	—	—	82.0	3,610.0	3,332.0	18.2	—	—	—	—	—	5.4	3.6	4.2
639	—	—	—	74.0	3,273.0	3,019.0	16.7	—	—	—	—	—	4.6	3.4	3.8
640	—	—	—	82.0	3,630.0	3,349.0	18.5	—	—	—	—	—	5.1	3.7	4.2
641	—	—	—	—	—	—	4.1	—	—	—	—	—	10.9	1.1	2.5
642	—	—	—	—	—	—	12.8	—	—	—	—	—	34.0	3.3	7.9
643	40.0	1.63	1.34	—	—	—	2.7	—	—	—	—	—	38.7	1.5	4.5
644	45.0	1.81	1.49	—	—	—	3.0	—	—	—	—	—	43.1	1.7	5.0
645	—	—	—	—	—	—	3.5	—	—	—	—	—	5.8	0.8	3.4
646	—	—	—	—	—	—	14.5	—	—	—	—	—	23.8	3.2	14.0
647	43.0	1.72	1.41	—	—	—	13.6	—	—	—	—	—	17.6	2.9	11.6
648	48.0	1.93	1.58	—	—	—	15.2	—	—	—	—	—	19.7	3.2	13.0
649	40.0	1.62	1.33	—	—	—	8.8	—	—	—	—	—	20.4	2.1	9.4
650	47.0	1.89	1.55	—	—	—	10.3	—	—	—	—	—	23.8	2.4	11.0
651	28.0	1.15	0.95	—	—	—	4.6	—	—	—	—	—	30.3	0.8	7.0
652	33.0	1.38	1.14	—	—	—	5.5	—	—	—	—	—	36.3	0.9	8.4
653	—	—	—	—	—	—	2.8	—	—	—	—	—	6.2	0.7	2.3
654	—	—	—	—	—	—	10.4	—	—	—	—	—	23.2	2.7	8.6
655	40.0	1.59	1.30	—	—	—	7.4	35.0	—	—	2.0	26.0	26.1	1.9	9.9
656	46.0	1.85	1.51	—	—	—	8.6	41.0	—	—	2.0	30.0	30.3	2.2	11.5

TABLE 1   Composition of Important Feeds: Energy Values, Proximate Analyses, Plant Cell Wall Constituents, and Acid Detergent Fiber—*Continued*

Entry Number	Feed Name Description	International Feed Number	Dry Matter (%)	Ruminants TDN (%)	DE (Mcal/ kg)	ME (Mcal/ kg)	NE$_m$ (Mcal/ kg)	NE$_g$ (Mcal/ kg)	Dairy Cattle NE$_l$ (Mcal/ kg)	Chickens ME$_n$ (kcal/ kg)	TME (kcal/ kg)	NE$_p$ (kcal/ kg)
	SAFFLOWER *Carthamus tinctorius*											
657	seeds	4-07-958	94.0	83.0	3.68	3.29	2.02	1.38	1.93	—	—	—
658			100.0	89.0	3.92	3.51	2.15	1.47	2.06	—	—	—
659	seeds, meal mech extd	5-04-109	91.0	55.0	2.42	2.03	1.20	0.60	1.24	—	—	—
660			100.0	60.0	2.65	2.22	1.31	0.65	1.35	—	—	—
661	seeds, meal solv extd	5-04-110	92.0	52.0	2.32	1.93	1.14	0.50	1.18	1,193.0	—	1,035.0
662			100.0	57.0	2.51	2.09	1.23	0.55	1.28	1,294.0	—	1,122.0
663	seeds without hulls, meal solv extd	5-07-959	92.0	67.0	2.95	2.57	1.53	0.97	1.53	1,921.0	—	1,185.0
664			100.0	73.0	3.22	2.80	1.67	1.06	1.67	2,096.0	—	1,293.0
	SAGE, BLACK *Salvia mellifera*											
665	browse, fresh, stem-cured	2-05-564	65.0	32.0	1.40	1.13	0.68	0.15	0.70	—	—	—
666			100.0	49.0	2.16	1.73	1.05	0.23	1.08	—	—	—
	SAGEBRUSH, BIG *Artemisia tridentata*											
667	browse, fresh, stem-cured	2-07-992	65.0	33.0	1.43	1.16	0.70	0.18	0.72	—	—	—
668			100.0	50.0	2.21	1.78	1.07	0.28	1.11	—	—	—
	SAGEBRUSH, BUD *Artemisia tridentata*											
669	browse, fresh, early vegetative	2-07-991	23.0	12.0	0.52	0.42	0.25	0.07	0.26	—	—	—
670			100.0	51.0	2.25	1.82	1.10	0.32	1.13	—	—	—
671	browse, fresh, late vegetative	2-04-124	32.0	17.0	0.73	0.60	0.36	0.11	0.37	—	—	—
672			100.0	52.0	2.29	1.87	1.12	0.36	1.15	—	—	—
	SAGEBRUSH, FRINGED *Artemisia frigida*											
673	browse, fresh, midbloom	2-04-129	43.0	25.0	1.10	0.92	0.54	0.25	0.56	—	—	—
674			100.0	58.0	2.56	2.13	1.26	0.58	1.30	—	—	—
675	browse, fresh, mature	2-04-130	60.0	30.0	1.35	1.09	0.66	0.19	0.68	—	—	—
676			100.0	51.0	2.25	1.82	1.10	0.32	1.13	—	—	—
	SALTBUSH, NUTTALL *Atriplex nuttallii*											
677	browse, fresh, stem-cured	2-07-993	55.0	20.0	0.87	0.63	0.46	—	0.42	—	—	—
678			100.0	36.0	1.59	1.15	0.84	—	0.76	—	—	—
	SALTGRASS *Distichlis* spp											
679	fresh, postripe	2-04-169	74.0	40.0	1.74	1.42	0.85	0.29	0.88	—	—	—
680			100.0	53.0	2.34	1.91	1.14	0.40	1.18	—	—	—
681	hay, sun-cured	1-04-168	89.0	45.0	2.01	1.63	0.98	0.28	1.01	—	—	—
682			100.0	51.0	2.25	1.82	1.10	0.32	1.13	—	—	—
	SALTGRASS, DESERT *Distichlis stricta*											
683	fresh	2-04-171	29.0	17.0	0.75	0.63	0.37	0.18	0.38	—	—	—
684			100.0	59.0	2.60	2.18	1.28	0.62	1.33	—	—	—
	SCREENINGS—SEE BARLEY, SEE CEREALS, SEE WHEAT											

	Horses			Swine				Plant Cell Wall Constituents							
Entry Num- ber	TDN (%)	DE (Mcal/ kg)	ME (Mcal/ kg)	TDN (%)	DE (kcal/ kg)	ME (kcal/ kg)	Crude Pro- tein (%)	Cell Walls (%)	Cell- ulose (%)	Hemi- cell- ulose (%)	Lig- nin (%)	Acid Deter- gent Fiber (%)	Crude Fiber (%)	Ether Ex- tract (%)	Ash (%)
657	—	—	—	—	—	—	16.3	—	—	—	—	—	26.8	32.8	2.9
658	—	—	—	—	—	—	17.4	—	—	—	—	—	28.6	35.1	3.1
659	—	—	—	—	1,525.0	1,396.0	20.2	54.0	—	—	—	38.0	32.4	6.1	3.8
660	—	—	—	—	1,667.0	1,526.0	22.1	59.0	—	—	—	41.0	35.4	6.7	4.1
661	—	—	—	—	—	—	23.4	53.0	—	—	13.0	38.0	30.0	1.4	5.4
662	—	—	—	—	—	—	25.4	58.0	—	—	14.0	41.0	32.5	1.5	5.9
663	—	—	—	—	3,361.0	2,908.0	43.0	—	—	—	—	—	13.5	1.3	7.5
664	—	—	—	—	3,666.0	3,172.0	46.9	—	—	—	—	—	14.7	1.4	8.2
665	—	—	—	—	—	—	5.5	—	—	—	—	—	—	7.0	3.6
666	—	—	—	—	—	—	8.5	—	—	—	—	—	—	10.8	5.5
667	—	—	—	—	—	—	6.1	27.0	—	—	8.0	20.0	—	7.2	4.3
668	—	—	—	—	—	—	9.3	42.0	—	—	12.0	30.0	—	11.0	6.6
669	—	—	—	—	—	—	4.0	—	—	—	—	—	—	1.1	4.9
670	—	—	—	—	—	—	17.3	—	—	—	—	—	—	4.9	21.4
671	—	—	—	—	—	—	5.6	—	—	—	—	—	7.3	0.8	6.9
672	—	—	—	—	—	—	17.5	—	—	—	—	—	22.7	2.5	21.6
673	—	—	—	—	—	—	4.0	—	—	—	—	—	14.3	0.9	2.8
674	—	—	—	—	—	—	9.4	—	—	—	—	—	33.2	2.0	6.5
675	—	—	—	—	—	—	4.3	27.0	—	—	6.0	21.0	19.1	2.0	10.3
676	—	—	—	—	—	—	7.1	46.0	—	—	10.0	35.0	31.8	3.4	17.1
677	—	—	—	—	—	—	4.0	—	—	—	—	—	—	1.2	11.8
678	—	—	—	—	—	—	7.2	—	—	—	—	—	—	2.2	21.5
679	—	—	—	—	—	—	3.1	—	—	—	—	—	26.0	1.9	5.4
680	—	—	—	—	—	—	4.2	—	—	—	—	—	34.9	2.6	7.3
681	41.0	1.65	1.35	—	—	—	8.0	—	—	—	—	—	28.3	1.8	11.4
682	46.0	1.85	1.51	—	—	—	8.9	—	—	—	—	—	31.6	2.1	12.7
683	—	—	—	—	—	—	1.7	—	—	—	—	—	8.6	0.5	2.0
684	—	—	—	—	—	—	5.9	—	—	—	—	—	29.7	1.7	6.8

TABLE 1   Composition of Important Feeds: Energy Values, Proximate Analyses, Plant Cell Wall Constituents, and Acid Detergent Fiber—*Continued*

Entry Number	Feed Name Description	International Feed Number	Dry Matter (%)	Ruminants TDN (%)	DE (Mcal/kg)	ME (Mcal/kg)	NE$_m$ (Mcal/kg)	NE$_g$ (Mcal/kg)	Dairy Cattle NE$_l$ (Mcal/kg)	Chickens ME$_n$ (kcal/kg)	TME (kcal/kg)	NE$_p$ (kcal/kg)
	SEAWEED, KELP *Laminariales* (order)–*Fucales* (order)											
685	whole, dehy	4-08-073	91.0	29.0	1.29	0.89	0.73	—	0.61	—	—	—
686			100.0	32.0	1.41	0.98	0.80	—	0.66	—	—	—
	SEDGE *Carex* spp											
687	hay, sun-cured	1-04-193	89.0	47.0	2.05	1.67	1.00	0.32	1.03	—	—	—
688			100.0	52.0	2.29	1.87	1.12	0.36	1.15	—	—	—
	SESAME *Sesamum indicum*											
689	seeds, meal mech extd	5-04-220	93.0	71.0	3.15	2.76	1.66	1.08	1.64	2,213.0	—	1,708.0
690			100.0	77.0	3.40	2.98	1.79	1.17	1.77	2,387.0	—	1,842.0
	SHRIMP *Pandalus* spp–*Penaeus* spp											
691	process residue, meal	5-04-226	90.0	41.0	1.83	1.44	0.90	0.09	0.91	1,920.0	—	883.0
692	(Shrimp meal)		100.0	46.0	2.03	1.60	0.99	0.10	1.01	2,131.0	—	980.0
	SOLKA FLOC											
693		1-28-258	93.0	65.0	2.87	2.48	1.47	0.90	1.48	—	—	—
694			100.0	70.0	3.09	2.67	1.58	0.97	1.60	—	—	—
	SORGHUM *Sorghum bicolor*											
695	aerial part with heads, sun-cured (Fodder)	1-07-960	89.0	52.0	2.28	1.90	1.12	0.52	1.16	—	—	—
696			100.0	58.0	2.56	2.13	1.26	0.58	1.30	—	—	—
697	aerial part without heads, sun-cured (Stover)	1-04-302	88.0	48.0	2.11	1.73	1.03	0.38	1.06	—	—	—
698			100.0	54.0	2.38	1.96	1.16	0.43	1.20	—	—	—
699	distillers grains, dehy	5-04-374	94.0	78.0	3.43	3.04	1.85	1.24	1.79	—	—	—
700			100.0	83.0	3.66	3.25	1.97	1.32	1.91	—	—	—
701	grain	4-04-383	90.0	78.0	3.40	3.03	1.85	1.25	1.78	3,311.0	3,359.0	2,564.0
702			100.0	86.0	3.79	3.38	2.06	1.40	1.99	3,691.0	3,745.0	2,858.0
703	grain, less than 8% protein	4-20-892	88.0	82.0	3.61	3.25	2.01	1.38	1.90	—	—	—
704			100.0	93.0	4.10	3.69	2.28	1.57	2.16	—	—	—
705	grain, 8–10% protein	4-20-893	87.0	80.0	3.52	3.17	1.95	1.34	1.85	3,288.0	—	—
706			100.0	92.0	4.06	3.65	2.25	1.54	2.13	3,786.0	—	—
707	grain, more than 10% protein	4-20-894	88.0	80.0	3.53	3.17	1.95	1.34	1.86	—	—	—
708			100.0	91.0	4.01	3.60	2.22	1.52	2.11	—	—	—
709	hay, sun-cured, early vegetative (South)	1-04-299	92.0	54.0	2.39	2.00	1.18	0.57	1.22	—	—	—
710			100.0	59.0	2.60	2.18	1.28	0.62	1.33	—	—	—
711	hay, sun-cured, late vegetative (South)	1-06-141	92.0	49.0	2.15	1.76	1.05	0.36	1.08	—	—	—
712			100.0	53.0	2.34	1.91	1.14	0.40	1.18	—	—	—
713	hay, sun-cured, early bloom (South)	1-06-142	93.0	47.0	2.05	1.65	1.00	0.26	1.03	—	—	—
714			100.0	50.0	2.21	1.78	1.07	0.28	1.11	—	—	—
715	silage	3-04-323	30.0	18.0	0.78	0.66	0.39	0.19	0.40	—	—	—
716			100.0	60.0	2.65	2.22	1.31	0.65	1.35	—	—	—
717	silage, dough stage	3-04-321	28.0	16.0	0.69	0.57	0.34	0.13	0.35	—	—	—
718			100.0	55.0	2.43	2.00	1.19	0.47	1.23	—	—	—
	SORGHUM, JOHNSONGRASS *Sorghum halepense*											
719	hay, sun-cured	1-04-407	89.0	48.0	2.09	1.71	1.02	0.35	1.05	—	—	—
720			100.0	53.0	2.34	1.91	1.14	0.40	1.18	—	—	—

	Horses			Swine				Plant Cell Wall Constituents							
Entry Num-ber	TDN (%)	DE (Mcal/ kg)	ME (Mcal/ kg)	TDN (%)	DE (kcal/ kg)	ME (kcal/ kg)	Crude Pro-tein (%)	Cell Walls (%)	Cell-ulose (%)	Hemi-cell-ulose (%)	Lig-nin (%)	Acid Deter-gent Fiber (%)	Crude Fiber (%)	Ether Ex-tract (%)	Ash (%)
685	—	—	—	—	—	—	6.5	—	—	—	—	—	6.5	0.5	35.2
686	—	—	—	—	—	—	7.1	—	—	—	—	—	7.1	0.5	38.6
687	41.0	1.65	1.35	—	—	—	8.4	—	—	—	—	—	28.0	2.1	6.4
688	46.0	1.85	1.52	—	—	—	9.4	—	—	—	—	—	31.3	2.4	7.2
689	—	—	—	70.0	3,101.0	2,945.0	45.5	16.0	—	—	2.0	16.0	5.7	6.9	11.2
690	—	—	—	76.0	3,344.0	3,176.0	49.1	17.0	—	—	2.0	17.0	6.1	7.5	12.1
691	—	—	—	—	2,420.0	2,108.0	39.9	—	—	—	—	15.0	14.1	3.9	26.8
692	—	—	—	—	2,686.0	2,339.0	44.2	—	—	—	—	17.0	15.6	4.3	29.7
693	—	—	—	—	—	—	—	—	—	—	—	—	—	—	—
694	—	—	—	—	—	—	—	—	—	—	—	—	—	—	—
695	41.0	1.64	1.34	—	—	—	6.7	—	—	—	—	—	23.9	2.2	8.4
696	46.0	1.84	1.51	—	—	—	7.5	—	—	—	—	—	26.9	2.4	9.4
697	35.0	1.43	1.17	—	—	—	4.6	—	—	—	—	—	29.6	1.5	9.7
698	40.0	1.62	1.33	—	—	—	5.2	—	—	—	—	—	33.5	1.7	11.0
699	—	—	—	—	—	—	32.2	—	—	—	—	—	11.9	8.9	3.6
700	—	—	—	—	—	—	34.4	—	—	—	—	—	12.7	9.5	3.8
701	—	—	—	78.0	3,431.0	3,216.0	11.1	21.0	3.0	16.0	—	5.0	2.4	2.8	1.8
702	—	—	—	87.0	3,824.0	3,585.0	12.4	23.0	3.0	18.0	—	5.0	2.6	3.1	2.0
703	—	—	—	—	—	—	6.8	—	—	—	—	—	—	—	—
704	—	—	—	—	—	—	7.7	—	—	—	—	—	—	—	—
705	—	—	—	74.0	3,270.0	3,199.0	8.8	—	—	—	—	—	2.3	2.9	1.8
706	—	—	—	85.0	3,766.0	3,684.0	10.1	—	—	—	—	—	2.6	3.4	2.1
707	—	—	—	—	—	—	11.0	—	—	—	—	—	—	—	—
708	—	—	—	—	—	—	12.5	—	—	—	—	—	—	—	—
709	—	—	—	—	—	—	14.7	60.0	25.0	—	4.0	30.0	25.8	3.0	12.0
710	—	—	—	—	—	—	16.0	65.0	27.0	—	4.0	33.0	28.0	3.3	13.0
711	—	—	—	—	—	—	11.0	64.0	29.0	—	5.0	36.0	30.4	2.4	10.1
712	—	—	—	—	—	—	12.0	70.0	31.0	—	5.0	39.0	33.0	2.6	11.0
713	—	—	—	—	—	—	7.0	70.0	33.0	—	6.0	42.0	35.3	1.9	8.4
714	—	—	—	—	—	—	7.5	75.0	35.0	—	7.0	45.0	38.0	2.0	9.0
715	—	—	—	16.0	722.0	682.0	2.2	—	—	—	2.0	11.0	8.2	0.9	2.6
716	—	—	—	55.0	2,441.0	2,306.0	7.5	—	—	—	6.0	38.0	27.9	3.0	8.7
717	—	—	—	—	—	—	1.7	—	—	—	—	—	8.1	0.9	2.6
718	—	—	—	—	—	—	6.0	—	—	—	—	—	28.5	3.3	9.3
719	41.0	1.66	1.36	—	—	—	8.5	—	—	—	—	—	29.9	2.1	7.3
720	46.0	1.86	1.52	—	—	—	9.5	—	—	—	—	—	33.5	2.4	8.2

TABLE 1 Composition of Important Feeds: Energy Values, Proximate Analyses, Plant Cell Wall Constituents, and Acid Detergent Fiber—*Continued*

Entry Number	Feed Name Description	International Feed Number	Dry Matter (%)	Ruminants TDN (%)	DE (Mcal/kg)	ME (Mcal/kg)	NE$_m$ (Mcal/kg)	NE$_g$ (Mcal/kg)	Dairy Cattle NE$_l$ (Mcal/kg)	Chickens ME$_n$ (kcal/kg)	TME (kcal/kg)	NE$_p$ (kcal/kg)
	SORGHUM, SORGO *Sorghum bicolor saccharatum*											
721	silage	3-04-468	27.0	16.0	0.70	0.59	0.35	0.16	0.36	—	—	—
722			100.0	58.0	2.56	2.13	1.26	0.58	1.30	—	—	—
	SORGHUM, SUDANGRASS *Sorghum bicolor sudanense*											
723	fresh, early vegetative	2-04-484	18.0	12.0	0.55	0.48	0.28	0.17	0.28	—	—	—
724			100.0	70.0	3.09	2.67	1.58	0.97	1.60	—	—	—
725	fresh, midbloom	2-04-485	23.0	14.0	0.63	0.54	0.32	0.17	0.32	—	—	—
726			100.0	63.0	2.78	2.36	1.39	0.75	1.42	—	—	—
727	hay, sun-cured, full bloom	1-04-480	91.0	51.0	2.25	1.86	1.10	0.46	1.14	—	—	—
728			100.0	56.0	2.47	2.04	1.21	0.51	1.25	—	—	—
729	silage	3-04-499	28.0	16.0	0.69	0.57	0.34	0.13	0.35	—	—	—
730			100.0	55.0	2.43	2.00	1.19	0.47	1.23	—	—	—
	SOYBEAN *Glycine max*											
731	flour by-product (Soybean	4-04-594	90.0	46.0	2.01	1.63	0.98	0.28	1.01	774.0	—	466.0
732	mill feed)		100.0	51.0	2.25	1.82	1.10	0.32	1.13	865.0	—	521.0
733	fresh, dough stage	2-04-573	26.0	17.0	0.75	0.64	0.38	0.22	0.39	—	—	—
734			100.0	66.0	2.91	2.49	1.47	0.85	1.50	—	—	—
735	hay, sun-cured, midbloom	1-04-538	94.0	50.0	2.19	1.79	1.07	0.37	1.10	—	—	—
736			100.0	53.0	2.34	1.91	1.14	0.40	1.18	—	—	—
737	hay, sun-cured, dough stage	1-04-542	88.0	54.0	2.36	1.99	1.17	0.60	1.21	—	—	—
738			100.0	61.0	2.69	2.27	1.33	0.69	1.38	—	—	—
739	hulls	1-04-560	91.0	70.0	3.09	2.71	1.63	1.06	1.61	668.0	—	—
740			100.0	77.0	3.40	2.98	1.79	1.17	1.77	734.0	—	—
	oil—see Fats and oils											
741	protein concentrate, more	5-08-038	92.0	70.0	3.08	2.69	1.61	1.04	1.60	2,472.0	—	1,798.0
742	than 70% protein		100.0	76.0	3.35	2.93	1.76	1.14	1.74	2,695.0	—	1,960.0
743	seeds	5-04-610	92.0	83.0	3.67	3.30	2.03	1.39	1.93	3,363.0	—	2,298.0
744			100.0	91.0	4.01	3.60	2.22	1.52	2.11	3,674.0	—	2,511.0
745	seeds, heat processed	5-04-597	90.0	84.0	3.73	3.36	2.08	1.43	1.97	—	—	—
746			100.0	94.0	4.14	3.74	2.31	1.59	2.18	—	—	—
747	seeds, meal mech extd	5-04-600	90.0	77.0	3.37	3.00	1.83	1.23	1.77	2,429.0	—	1,722.0
748			100.0	85.0	3.75	3.34	2.03	1.37	1.96	2,699.0	—	1,914.0
749	seeds, meal solv extd,	5-20-637	89.0	75.0	3.31	2.94	1.79	1.20	1.73	2,219.0	2,639.0	1,589.0
750	44% protein		100.0	84.0	3.70	3.29	2.00	1.35	1.94	2,485.0	2,956.0	1,779.0
751	seeds without hulls, meal	5-04-612	90.0	78.0	3.46	3.09	1.89	1.28	1.81	2,455.0	—	1,658.0
752	solv extd		100.0	87.0	3.84	3.42	2.09	1.42	2.01	2,724.0	—	1,839.0
753	silage	3-04-581	27.0	15.0	0.66	0.55	0.33	0.13	0.34	—	—	—
754			100.0	55.0	2.43	2.00	1.19	0.47	1.23	—	—	—
755	straw	1-04-567	88.0	36.0	1.62	1.25	0.81	—	0.80	—	—	—
756			100.0	42.0	1.85	1.42	0.92	—	0.91	—	—	—
	SPELT *Triticum spelta*											
757	grain	4-04-651	90.0	67.0	2.97	2.60	1.55	1.00	1.54	—	—	—
758			100.0	75.0	3.31	2.89	1.73	1.11	1.72	—	—	—
	SQUIRRELTAIL *Sitanion* spp											
759	fresh, stem-cured	2-05-566	50.0	25.0	1.10	0.89	0.54	0.14	0.55	—	—	—
760			100.0	50.0	2.21	1.78	1.07	0.28	1.11	—	—	—
	SUDANGRASS—SEE SORGHUM, SUDANGRASS											

| | Horses | | | Swine | | | | Plant Cell Wall Constituents | | | | | | | |
| | | | | | | | Crude Pro-tein (%) | Cell Walls (%) | Cell-ulose (%) | Hemi-cell-ulose (%) | Lig-nin (%) | Acid Deter-gent Fiber (%) | Crude Fiber (%) | Ether Ex-tract (%) | Ash (%) |
Entry Num-ber	TDN (%)	DE (Mcal/ kg)	ME (Mcal/ kg)	TDN (%)	DE (kcal/ kg)	ME (kcal/ kg)									
721	—	—	—	—	—	—	1.7	—	—	—	—	—	7.8	0.7	1.8
722	—	—	—	—	—	—	6.2	—	—	—	—	—	28.3	2.6	6.4
723	—	—	—	—	—	—	3.0	10.0	5.0	4.0	1.0	5.0	4.1	0.7	1.6
724	—	—	—	—	—	—	16.8	55.0	26.0	24.0	3.0	29.0	23.0	3.9	9.0
725	—	—	—	14.0	619.0	584.0	2.0	15.0	8.0	6.0	1.0	9.0	6.8	0.4	2.4
726	—	—	—	62.0	2,719.0	2,562.0	8.8	65.0	34.0	25.0	5.0	40.0	30.0	1.8	10.5
727	42.0	1.69	1.38	—	—	—	7.3	62.0	32.0	24.0	5.0	38.0	32.8	1.6	8.7
728	46.0	1.85	1.52	—	—	—	8.0	68.0	35.0	26.0	6.0	42.0	36.0	1.8	9.6
729	—	—	—	—	—	—	3.1	—	11.0	—	1.0	12.0	9.4	0.8	2.8
730	—	—	—	—	—	—	10.8	—	38.0	—	5.0	42.0	33.1	2.8	9.8
731	—	—	—	26.0	1,167.0	925.0	12.6	—	—	—	—	—	34.1	1.8	4.9
732	—	—	—	30.0	1,303.0	1,032.0	14.0	—	—	—	—	—	38.1	2.0	5.5
733	—	—	—	—	—	—	4.5	—	—	—	—	—	7.4	1.3	2.5
734	—	—	—	—	—	—	17.7	—	—	—	—	—	29.0	5.1	9.8
735	25.0	1.06	0.87	—	—	—	16.7	—	—	—	—	—	27.9	5.1	8.3
736	26.0	1.14	0.93	—	—	—	17.8	—	—	—	—	—	29.8	5.4	8.8
737	37.0	1.50	1.23	—	—	—	14.7	—	—	—	—	—	25.0	3.6	6.0
738	42.0	1.71	1.40	—	—	—	16.8	—	—	—	—	—	28.5	4.1	6.8
739	41.0	1.64	1.35	47.0	1,887.0	1,765.0	11.0	61.0	42.0	16.0	2.0	45.0	36.4	1.9	4.6
740	45.0	1.81	1.48	52.0	2,074.0	1,940.0	12.1	67.0	46.0	18.0	2.0	50.0	40.1	2.1	5.1
741	—	—	—	—	4,934.0	3,821.0	84.3	—	—	—	—	—	0.1	0.5	3.5
742	—	—	—	—	5,377.0	4,164.0	91.9	—	—	—	—	—	0.1	0.6	3.8
743	—	—	—	93.0	4,092.0	3,574.0	39.2	—	—	—	—	9.0	5.3	17.2	5.1
744	—	—	—	101.0	4,471.0	3,905.0	42.8	—	—	—	—	10.0	5.8	18.8	5.5
745	—	—	—	92.0	4,056.0	3,540.0	38.0	—	—	—	—	10.0	5.0	18.0	4.6
746	—	—	—	102.0	4,507.0	3,933.0	42.2	—	—	—	—	11.0	5.6	20.0	5.1
747	—	—	—	79.0	3,610.0	2,972.0	42.9	—	—	—	—	—	5.9	4.8	6.0
748	—	—	—	88.0	4,013.0	3,304.0	47.7	—	—	—	—	—	6.6	5.3	6.7
749	—	—	—	75.0	3,318.0	2,817.0	44.6	—	—	—	—	—	6.2	1.4	6.5
750	—	—	—	84.0	3,716.0	3,155.0	49.9	—	—	—	—	—	7.0	1.5	7.3
751	—	—	—	76.0	3,942.0	3,155.0	49.7	—	—	—	—	—	3.4	0.9	5.8
752	—	—	—	84.0	4,373.0	3,500.0	55.1	—	—	—	—	—	3.7	1.0	6.5
753	—	—	—	—	—	—	4.8	—	—	—	—	—	7.8	0.7	2.7
754	—	—	—	—	—	—	17.3	—	—	—	—	—	28.4	2.7	9.7
755	30.0	1.25	1.02	—	—	—	4.6	61.0	33.0	—	14.0	47.0	38.9	1.3	5.6
756	34.0	1.42	1.17	—	—	—	5.2	70.0	38.0	—	16.0	54.0	44.3	1.5	6.4
757	—	—	—	70.0	3,070.0	2,865.0	12.0	—	—	—	—	—	9.1	1.9	3.5
758	—	—	—	77.0	3,416.0	3,187.0	13.3	—	—	—	—	—	10.2	2.1	3.9
759	—	—	—	—	—	—	1.6	—	—	—	—	—	—	1.1	8.5
760	—	—	—	—	—	—	3.1	—	—	—	—	—	—	2.2	17.0

TABLE 1   Composition of Important Feeds: Energy Values, Proximate Analyses, Plant Cell Wall Constituents, and Acid Detergent Fiber—*Continued*

Entry Number	Feed Name Description	International Feed Number	Dry Matter (%)	Ruminants TDN (%)	DE (Mcal/kg)	ME (Mcal/kg)	NE$_m$ (Mcal/kg)	NE$_g$ (Mcal/kg)	Dairy Cattle NE$_l$ (Mcal/kg)	Chickens ME$_n$ (kcal/kg)	TME (kcal/kg)	NE$_p$ (kcal/kg)
	SUGARCANE *Saccharum officinarum*											
761	bagasse, dehy	1-04-686	91.0	44.0	1.93	1.54	0.94	0.17	0.96	—	—	—
762			100.0	48.0	2.12	1.69	1.03	0.19	1.06	—	—	—
	molasses—see Molasses and syrup											
763	stems, fresh	2-13-248	15.0	9.0	0.41	0.34	0.20	0.10	0.21	—	—	—
764			100.0	61.0	2.69	2.27	1.33	0.69	1.38	—	—	—
765	sugar	4-04-701	100.0	98.0	4.30	3.90	2.43	1.68	2.27	3,715.0	—	2,794.0
766			100.0	98.0	4.32	3.91	2.44	1.68	2.28	3,729.0	—	2,805.0
	SUNFLOWER, COMMON *Helianthus annuus*											
767	seeds, meal solv extd	5-09-340	90.0	40.0	1.75	1.36	0.86	0.01	0.86	1,543.0	—	—
768			100.0	44.0	1.94	1.51	0.96	0.01	0.96	1,715.0	—	—
769	seeds without hulls, meal	5-04-738	93.0	69.0	3.03	2.64	1.57	1.00	1.57	2,218.0	—	1,220.0
770	mech extd		100.0	74.0	3.26	2.85	1.70	1.08	1.69	2,391.0	—	1,315.0
771	seeds without hulls, meal	5-04-739	93.0	60.0	2.67	2.27	1.34	0.76	1.37	2,085.0	2,390.0	1,213.0
772	solv extd		100.0	65.0	2.87	2.45	1.44	0.82	1.47	2,242.0	2,570.0	1,304.0
	SWEETCLOVER, YELLOW *Melilotus officinalis*											
773	hay, sun-cured	1-04-754	87.0	47.0	2.08	1.71	1.01	0.38	1.05	—	—	—
774			100.0	54.0	2.38	1.96	1.16	0.43	1.20	—	—	—
	SWINE *Sus scrofa*											
	fat—see Fats and oils											
775	livers, fresh	5-04-792	30.0	29.0	1.29	1.17	0.73	0.50	0.68	—	—	—
776			100.0	97.0	4.28	3.87	2.40	1.66	2.26	—	—	—
777	lungs, fresh	5-26-140	16.0	15.0	0.65	0.58	0.36	0.25	0.34	—	—	—
778			100.0	93.0	4.10	3.69	2.28	1.57	2.16	—	—	—
	TIMOTHY *Phleum pratense*											
779	fresh, late vegetative	2-04-903	26.0	19.0	0.84	0.73	0.43	0.27	0.43	—	—	—
780			100.0	72.0	3.17	2.76	1.64	1.03	1.64	—	—	—
781	fresh, midbloom	2-04-905	29.0	18.0	0.81	0.69	0.41	0.22	0.42	—	—	—
782			100.0	63.0	2.78	2.36	1.39	0.75	1.42	—	—	—
783	hay, sun-cured, late vege-	1-04-881	89.0	63.0	2.80	2.42	1.44	0.89	1.45	—	—	—
784	tative		100.0	71.0	3.13	2.71	1.61	1.00	1.62	—	—	—
785	hay, sun-cured, early bloom	1-04-882	90.0	59.0	2.61	2.23	1.31	0.76	1.34	—	—	—
786			100.0	66.0	2.91	2.49	1.47	0.85	1.50	—	—	—
787	hay, sun-cured, midbloom	1-04-883	89.0	54.0	2.39	2.01	1.19	0.61	1.22	—	—	—
788			100.0	61.0	2.69	2.27	1.33	0.69	1.38	—	—	—
789	hay, sun-cured, full bloom	1-04-884	89.0	52.0	2.27	1.89	1.12	0.52	1.15	—	—	—
790			100.0	58.0	2.56	2.13	1.26	0.58	1.30	—	—	—
791	hay, sun-cured, late bloom	1-04-885	88.0	50.0	2.18	1.81	1.07	0.45	1.11	—	—	—
792			100.0	56.0	2.47	2.04	1.21	0.51	1.25	—	—	—
793	hay, sun-cured, milk stage	1-04-886	92.0	47.0	2.10	1.71	1.02	0.33	1.06	—	—	—
794			100.0	52.0	2.29	1.87	1.12	0.36	1.15	—	—	—
795	silage, early bloom	3-04-918	36.0	22.0	0.96	0.81	0.48	0.24	0.49	—	—	—
796			100.0	60.0	2.65	2.22	1.31	0.65	1.35	—	—	—
797	silage, full bloom	3-04-920	36.0	21.0	0.93	0.78	0.46	0.22	0.47	—	—	—
798			100.0	59.0	2.60	2.18	1.28	0.62	1.33	—	—	—
799	silage, milk stage	3-04-921	42.0	23.0	1.03	0.85	0.50	0.21	0.52	—	—	—
800			100.0	56.0	2.47	2.04	1.21	0.51	1.25	—	—	—

| Entry Num-ber | Horses | | | Swine | | | Crude Pro-tein (%) | Plant Cell Wall Constituents | | | | | Crude Fiber (%) | Ether Ex-tract (%) | Ash (%) |
	TDN (%)	DE (Mcal/ kg)	ME (Mcal/ kg)	TDN (%)	DE (kcal/ kg)	ME (kcal/ kg)		Cell Walls (%)	Cell-ulose (%)	Hemi-cell-ulose (%)	Lig-nin (%)	Acid Deter-gent Fiber (%)			
761	—	—	—	—	—	—	1.5	—	—	—	—	—	43.9	0.7	2.9
762	—	—	—	—	—	—	1.6	—	—	—	—	—	48.1	0.7	3.2
763	—	—	—	—	—	—	1.2	11.0	5.0	4.0	2.0	—	4.2	0.1	0.9
764	—	—	—	—	—	—	7.6	74.0	34.0	30.0	11.0	—	27.5	0.7	6.0
765	—	—	—	98.0	3,741.0	3,659.0	—	—	—	—	—	—	—	—	0.1
766	—	—	—	99.0	3,755.0	3,673.0	—	—	—	—	—	—	—	—	0.1
767	—	—	—	—	1,991.0	1,807.0	23.3	36.0	—	—	11.0	30.0	31.6	1.1	5.6
768	—	—	—	—	2,213.0	2,009.0	25.9	40.0	—	—	12.0	33.0	35.1	1.2	6.3
769	—	—	—	71.0	3,113.0	2,737.0	41.4	—	—	—	—	—	12.2	8.0	6.6
770	—	—	—	77.0	3,357.0	2,951.0	44.6	—	—	—	—	—	13.1	8.7	7.1
771	—	—	—	69.0	3,049.0	2,652.0	46.3	—	—	—	—	—	11.4	2.9	7.6
772	—	—	—	74.0	3,278.0	2,851.0	49.8	—	—	—	—	—	12.2	3.1	8.1
773	44.0	1.75	1.44	—	—	—	13.7	—	—	—	—	—	29.2	1.7	7.7
774	50.0	2.01	1.64	—	—	—	15.7	—	—	—	—	—	33.4	2.0	8.8
775	—	—	—	—	—	—	20.8	—	—	—	—	—	0.1	5.0	1.6
776	—	—	—	—	—	—	68.8	—	—	—	—	—	0.3	16.5	5.3
777	—	—	—	—	—	—	14.0	—	—	—	—	—	0.1	2.5	0.8
778	—	—	—	—	—	—	88.6	—	—	—	—	—	0.3	15.8	5.1
779	—	—	—	—	—	—	4.8	—	—	—	—	—	8.5	1.0	1.8
780	—	—	—	—	—	—	18.0	—	—	—	—	—	32.1	3.8	6.6
781	—	—	—	—	—	—	2.7	19.0	9.0	—	1.0	11.0	9.8	0.9	1.9
782	—	—	—	—	—	—	9.1	64.0	31.0	—	4.0	37.0	33.5	3.0	6.6
783	56.0	2.21	1.82	—	—	—	15.2	49.0	25.0	23.0	3.0	26.0	24.1	2.5	6.3
784	63.0	2.48	2.03	—	—	—	17.0	55.0	28.0	26.0	3.0	29.0	27.0	2.8	7.1
785	52.0	2.05	1.68	—	—	—	13.4	55.0	28.0	26.0	4.0	29.0	25.1	2.6	5.1
786	58.0	2.29	1.87	—	—	—	15.0	61.0	31.0	29.0	4.0	32.0	28.0	2.9	5.7
787	51.0	2.00	1.64	—	—	—	8.1	60.0	29.0	28.0	4.0	32.0	27.6	2.3	5.6
788	57.0	2.25	1.84	—	—	—	9.1	67.0	33.0	31.0	5.0	36.0	31.0	2.6	6.3
789	39.0	1.58	1.29	—	—	—	7.2	60.0	30.0	27.0	5.0	34.0	28.4	2.7	4.6
790	44.0	1.78	1.46	—	—	—	8.1	68.0	34.0	30.0	6.0	38.0	32.0	3.1	5.2
791	42.0	1.67	1.37	—	—	—	6.9	62.0	30.0	26.0	6.0	35.0	28.7	2.5	4.8
792	47.0	1.89	1.55	—	—	—	7.8	70.0	34.0	29.0	7.0	40.0	32.5	2.8	5.4
793	41.0	1.65	1.35	—	—	—	6.4	65.0	—	27.0	7.0	38.0	31.0	2.1	5.8
794	45.0	1.80	1.47	—	—	—	7.0	71.0	—	30.0	8.0	41.0	33.9	2.3	6.3
795	—	—	—	—	—	—	3.7	—	—	—	—	—	12.8	1.2	2.5
796	—	—	—	—	—	—	10.2	—	—	—	—	—	35.3	3.2	6.8
797	—	—	—	—	—	—	3.5	—	—	—	—	—	12.9	1.2	2.5
798	—	—	—	—	—	—	9.7	—	—	—	—	—	36.3	3.2	6.9
799	—	—	—	—	—	—	3.5	23.0	—	5.0	—	21.0	15.9	1.3	2.7
800	—	—	—	—	—	—	8.4	55.0	—	11.0	—	51.0	38.4	3.1	6.6

TABLE 1 Composition of Important Feeds: Energy Values, Proximate Analyses, Plant Cell Wall Constituents, and Acid Detergent Fiber—*Continued*

Entry Number	Feed Name Description	International Feed Number	Dry Matter (%)	Ruminants TDN (%)	DE (Mcal/kg)	ME (Mcal/kg)	NE$_m$ (Mcal/kg)	NE$_g$ (Mcal/kg)	Dairy Cattle NE$_l$ (Mcal/kg)	Chickens ME$_n$ (kcal/kg)	TME (kcal/kg)	NE$_p$ (kcal/kg)
	TOMATO *Lycopersicon esculentum*											
801	pomace, dehy	5-05-041	92.0	53.0	2.35	1.96	1.16	0.53	1.19	1,751.0	—	—
802			100.0	58.0	2.56	2.13	1.26	0.58	1.30	1,908.0	—	—
	TORULA DRIED YEAST— SEE YEAST, TORULA											
	TREFOIL, BIRDSFOOT *Lotus corniculatus*											
803	fresh	2-20-786	24.0	16.0	0.71	0.60	0.36	0.21	0.36	—	—	—
804			100.0	66.0	2.91	2.49	1.47	0.85	1.50	—	—	—
805	hay, sun-cured	1-05-044	92.0	54.0	2.40	2.01	1.18	0.57	1.22	—	—	—
806			100.0	59.0	2.60	2.18	1.28	0.62	1.33	—	—	—
	TRITICALE *Triticale hexaploide*											
807	grain	4-20-362	90.0	76.0	3.33	2.96	1.80	1.21	1.74	3,163.0	3,256.0	2,200.0
808			100.0	84.0	3.70	3.29	2.00	1.35	1.94	3,521.0	3,625.0	2,450.0
	TURNIP *Brassica rapa rapa*											
809	roots, fresh	4-05-067	9.0	8.0	0.35	0.31	0.19	0.13	0.18	—	—	—
810			100.0	85.0	3.75	3.34	2.03	1.37	1.96	—	—	—
	UREA											
811	45% nitrogen, 281% protein	5-05-070	99.0	0.0	0.0	0.0	0.0	0.0	0.0	—	—	—
812	equivalent		100.0	0.0	0.0	0.0	0.0	0.0	0.0	—	—	—
	VETCH *Vicia* spp											
813	hay, sun-cured	1-05-106	89.0	51.0	2.24	1.86	1.10	0.49	1.14	—	—	—
814			100.0	57.0	2.51	2.09	1.23	0.55	1.28	—	—	—
	WHALE *Balaena glacialis– Balaenoptera* spp											
815	meat, meal rendered	5-05-160	91.0	80.0	3.55	3.17	1.94	1.32	1.86	—	—	—
816			100.0	88.0	3.88	3.47	2.12	1.45	2.04	—	—	—
	WHEAT *Triticum aestivum*											
817	bran	4-05-190	89.0	63.0	2.74	2.37	1.40	0.86	1.42	1,237.0	1,706.0	981.0
818			100.0	70.0	3.09	2.67	1.58	0.97	1.60	1,393.0	1,921.0	1,105.0
819	bread, dehy	4-07-944	95.0	82.0	3.61	3.22	1.96	1.33	1.89	3,268.0	—	—
820			100.0	86.0	3.79	3.38	2.06	1.40	1.99	3,429.0	—	—
821	flour, hard red spring, less	4-08-113	88.0	77.0	3.39	3.03	1.85	1.26	1.78	—	—	1,940.0
822	than 1.5% fiber		100.0	87.0	3.84	3.42	2.09	1.42	2.01	—	—	2,192.0
823	flour, less than 1.5% fiber	4-05-199	88.0	77.0	3.40	3.04	1.86	1.26	1.78	2,954.0	—	1,949.0
824	(Wheat feed flour)		100.0	88.0	3.88	3.47	2.12	1.45	2.04	3,375.0	—	2,227.0
825	flour, less than 2% fiber	4-28-221	88.0	76.0	3.34	2.97	1.81	1.23	1.75	—	1,954.0	—
826	(Feed flour)		100.0	86.0	3.79	3.38	2.06	1.40	1.99	—	2,220.0	—
827	flour by-product, less than	4-05-203	88.0	72.0	3.18	2.82	1.71	1.14	1.66	2,568.0	—	1,774.0
828	4% fiber (Wheat red dog)		100.0	82.0	3.62	3.20	1.94	1.30	1.89	2,916.0	—	2,015.0
829	flour by-product, less than	4-28-220	88.0	73.0	3.22	2.86	1.73	1.16	1.68	2,543.0	—	1,751.0
830	4.5% fiber (Middlings)		100.0	83.0	3.66	3.25	1.97	1.32	1.91	2,890.0	—	1,990.0
831	flour by-product, less than	4-05-201	88.0	65.0	2.85	2.48	1.48	0.93	1.48	2,162.0	2,544.0	1,417.0
832	7% fiber (Wheat shorts)		100.0	73.0	3.22	2.80	1.67	1.06	1.67	2,446.0	2,877.0	1,602.0

| | Horses | | | Swine | | | | Plant Cell Wall Constituents | | | | | | | |
Entry Number	TDN (%)	DE (Mcal/ kg)	ME (Mcal/ kg)	TDN (%)	DE (kcal/ kg)	ME (kcal/ kg)	Crude Protein (%)	Cell Walls (%)	Cellulose (%)	Hemicellulose (%)	Lignin (%)	Acid Detergent Fiber (%)	Crude Fiber (%)	Ether Extract (%)	Ash (%)
801	—	—	—	—	—	—	21.6	50.0	—	—	10.0	46.0	24.2	9.5	6.9
802	—	—	—	—	—	—	23.5	55.0	—	—	11.0	50.0	26.4	10.3	7.5
803	—	—	—	—	—	—	5.1	—	—	—	—	—	6.0	0.7	2.2
804	—	—	—	—	—	—	21.0	—	—	—	—	—	24.7	2.7	9.0
805	46.0	1.82	1.49	—	—	—	15.0	43.0	22.0	—	8.0	33.0	28.3	2.3	6.5
806	49.0	1.98	1.62	—	—	—	16.3	47.0	24.0	—	9.0	36.0	30.7	2.5	7.0
807	—	—	—	75.0	3,299.0	3,050.0	15.8	—	—	—	—	—	4.0	1.5	1.8
808	—	—	—	83.0	3,673.0	3,396.0	17.6	—	—	—	—	—	4.4	1.7	2.0
809	—	—	—	7.0	327.0	306.0	1.1	4.0	—	—	1.0	3.0	1.1	0.2	0.8
810	—	—	—	80.0	3,514.0	3,289.0	11.8	44.0	—	—	10.0	34.0	11.5	1.9	8.9
811	—	—	—	—	—	—	275.8	—	—	—	—	—	—	—	—
812	—	—	—	—	—	—	279.6	—	—	—	—	—	—	—	—
813	—	—	—	—	—	—	18.5	43.0	—	—	7.0	30.0	27.3	2.7	8.1
814	—	—	—	—	—	—	20.8	48.0	—	—	8.0	33.0	30.6	3.0	9.1
815	—	—	—	93.0	4,103.0	3,292.0	71.4	—	—	—	—	—	2.8	7.6	4.0
816	—	—	—	102.0	4,488.0	3,601.0	78.1	—	—	—	—	—	3.0	8.4	4.4
817	44.0	—	—	57.0	2,414.0	2,212.0	15.2	46.0	9.0	30.0	3.0	14.0	10.0	3.9	6.1
818	50.0	—	—	64.0	2,718.0	2,491.0	17.1	51.0	11.0	34.0	3.0	15.0	11.3	4.4	6.9
819	—	—	—	82.0	3,602.0	3,363.0	12.4	—	—	—	—	—	0.3	2.3	2.3
820	—	—	—	86.0	3,779.0	3,529.0	13.0	—	—	—	—	—	0.3	2.4	2.4
821	—	—	—	77.0	3,406.0	3,177.0	12.0	—	—	—	—	—	1.8	1.3	0.4
822	—	—	—	87.0	3,850.0	3,591.0	13.5	—	—	—	—	—	2.0	1.4	0.5
823	—	—	—	76.0	3,633.0	3,389.0	11.7	—	—	—	—	—	1.3	1.2	0.5
824	—	—	—	87.0	4,151.0	3,873.0	13.4	—	—	—	—	—	1.5	1.4	0.5
825	—	—	—	75.0	3,388.0	3,256.0	11.0	—	—	—	—	—	1.8	1.2	0.4
826	—	—	—	85.0	3,850.0	3,700.0	12.5	—	—	—	—	—	2.0	1.4	0.5
827	—	—	—	72.0	3,144.0	2,872.0	15.3	—	—	—	—	—	2.6	3.3	2.2
828	—	—	—	82.0	3,570.0	3,261.0	17.4	—	—	—	—	—	2.9	3.8	2.5
829	—	—	—	71.0	3,080.0	2,860.0	15.1	—	—	—	—	—	2.6	3.2	2.4
830	—	—	—	81.0	3,500.0	3,250.0	17.2	—	—	—	—	—	3.0	3.6	2.7
831	—	—	—	71.0	3,136.0	2,870.0	16.5	—	—	—	—	—	6.8	4.6	4.3
832	—	—	—	80.0	3,547.0	3,246.0	18.6	—	—	—	—	—	7.7	5.2	4.9

TABLE 1 Composition of Important Feeds: Energy Values, Proximate Analyses, Plant Cell Wall Constituents, and Acid Detergent Fiber—*Continued*

Entry Number	Feed Name Description	International Feed Number	Dry Matter (%)	Ruminants TDN (%)	DE (Mcal/kg)	ME (Mcal/kg)	NE$_m$ (Mcal/kg)	NE$_g$ (Mcal/kg)	Dairy Cattle NE$_l$ (Mcal/kg)	Chickens ME$_n$ (kcal/kg)	TME (kcal/kg)	NE$_p$ (kcal/kg)
833	flour by-product, less than	4-28-219	88.0	63.0	2.79	2.43	1.44	0.90	1.45	2,121.0	2,499.0	1,399.0
834	8% fiber (Shorts)		100.0	72.0	3.17	2.76	1.64	1.03	1.64	2,410.0	2,840.0	1,590.0
835	flour by-product, less than	4-05-205	89.0	61.0	2.70	2.33	1.38	0.83	1.40	2,115.0	—	1,496.0
836	9.5% fiber (Wheat mid-dlings)		100.0	69.0	3.04	2.62	1.55	0.94	1.57	2,380.0	—	1,684.0
837	fresh, early vegetative	2-05-176	22.0	16.0	0.72	0.62	0.37	0.23	0.37	—	—	—
838			100.0	73.0	3.22	2.80	1.67	1.06	1.67	—	—	—
839	germs, ground	5-05-218	88.0	83.0	3.66	3.30	2.04	1.40	1.93	2,694.0	—	1,619.0
840			100.0	94.0	4.14	3.74	2.31	1.59	2.18	3,051.0	—	1,834.0
841	grain	4-05-211	89.0	78.0	3.45	3.08	1.89	1.28	1.81	3,023.0	3,455.0	2,197.0
842			100.0	88.0	3.88	3.47	2.12	1.45	2.04	3,401.0	3,887.0	2,471.0
843	grain, hard red spring	4-05-258	88.0	78.0	3.44	3.08	1.89	1.29	1.81	2,702.0	3,408.0	1,931.0
844			100.0	89.0	3.92	3.51	2.15	1.47	2.06	3,084.0	3,890.0	2,205.0
845	grain, hard red winter	4-05-268	88.0	78.0	3.42	3.06	1.87	1.28	1.80	3,194.0	—	2,063.0
846			100.0	88.0	3.88	3.47	2.12	1.45	2.04	3,620.0	—	2,337.0
847	grain, soft red winter	4-05-294	88.0	78.0	3.46	3.10	1.90	1.30	1.82	3,089.0	—	1,703.0
848			100.0	89.0	3.92	3.51	2.15	1.47	2.06	3,500.0	—	1,929.0
849	grain, soft white winter	4-05-337	89.0	79.0	3.50	3.13	1.92	1.31	1.84	3,018.0	3,470.0	2,117.0
850			100.0	89.0	3.92	3.51	2.15	1.47	2.06	3,388.0	3,895.0	2,376.0
851	grain, soft white winter,	4-08-555	89.0	79.0	3.46	3.10	1.90	1.29	1.82	3,188.0	3,477.0	2,090.0
852	Pacific Coast		100.0	88.0	3.88	3.47	2.12	1.45	2.04	3,572.0	3,895.0	2,341.0
853	grain screenings	4-05-216	89.0	63.0	2.79	2.42	1.44	0.89	1.45	2,825.0	—	2,063.0
854			100.0	71.0	3.13	2.71	1.61	1.00	1.61	3,165.0	—	2,312.0
855	grits	4-07-852	90.0	79.0	3.48	3.11	1.91	1.30	1.83	—	—	—
856			100.0	88.0	3.88	3.47	2.12	1.45	2.04	—	—	—
857	hay, sun-cured	1-05-172	88.0	51.0	2.24	1.87	1.10	0.51	1.14	—	—	—
858			100.0	58.0	2.56	2.13	1.26	0.58	1.30	—	—	—
859	mill run, less than 9.5%	4-05-206	90.0	71.0	3.13	2.76	1.66	1.10	1.63	1,771.0	—	1,262.0
860	fiber		100.0	79.0	3.48	3.07	1.85	1.22	1.82	1,971.0	—	1,404.0
861	silage, early vegetative	3-05-184	30.0	17.0	0.75	0.62	0.37	0.16	0.38	—	—	—
862			100.0	57.0	2.51	2.09	1.23	0.55	1.28	—	—	—
863	silage, full bloom	3-05-185	25.0	15.0	0.65	0.55	0.32	0.16	0.33	—	—	—
864			100.0	59.0	2.60	2.18	1.28	0.62	1.33	—	—	—
865	straw	1-05-175	89.0	39.0	1.72	1.34	0.85	0.01	0.85	—	—	—
866			100.0	44.0	1.94	1.51	0.96	0.01	0.96	—	—	—
	WHEAT, DURUM *Triticum durum*											
867	grain	4-05-224	88.0	75.0	3.29	2.93	1.78	1.20	1.72	3,203.0	3,517.0	—
868			100.0	85.0	3.75	3.34	2.03	1.37	1.96	3,652.0	4,010.0	—
	WHEATGRASS, CRESTED *Agropyron desertorum*											
869	fresh, early vegetative	2-05-420	28.0	21.0	0.92	0.81	0.48	0.31	0.48	—	—	—
870			100.0	75.0	3.31	2.89	1.73	1.11	1.72	—	—	—
871	fresh, full bloom	2-05-424	45.0	27.0	1.21	1.02	0.60	0.31	0.62	—	—	—
872			100.0	61.0	2.69	2.27	1.33	0.69	1.38	—	—	—
873	fresh, postripe	2-05-428	80.0	39.0	1.73	1.39	0.84	0.19	0.86	—	—	—
874			100.0	49.0	2.16	1.73	1.05	0.23	1.08	—	—	—
875	hay, sun-cured	1-05-418	93.0	49.0	2.17	1.77	1.06	0.37	1.09	—	—	—
876			100.0	53.0	2.34	1.91	1.14	0.40	1.18	—	—	—
	WHEY											
877	dehy (Cattle)	4-01-182	93.0	75.0	3.33	2.95	1.78	1.19	1.74	1,949.0	1,662.0	1,548.0
878			100.0	81.0	3.57	3.16	1.91	1.27	1.87	2,087.0	1,780.0	1,659.0

	Horses			Swine				Plant Cell Wall Constituents							
Entry Number	TDN (%)	DE (Mcal/ kg)	ME (Mcal/ kg)	TDN (%)	DE (kcal/ kg)	ME (kcal/ kg)	Crude Protein (%)	Cell Walls (%)	Cellulose (%)	Hemicellulose (%)	Lignin (%)	Acid Detergent Fiber (%)	Crude Fiber (%)	Ether Extract (%)	Ash (%)
833	—	—	—	70.0	3,115.0	2,834.0	16.0	—	—	—	—	—	6.0	4.3	4.5
834	—	—	—	79.0	3,540.0	3,220.0	18.2	—	—	—	—	—	6.8	4.9	5.1
835	—	—	—	68.0	2,914.0	2,728.0	16.4	—	—	—	—	—	7.3	4.3	4.7
836	—	—	—	77.0	3,279.0	3,070.0	18.4	—	—	—	—	—	8.2	4.9	5.2
837	—	—	—	—	—	—	6.3	12.0	—	—	1.0	7.0	3.9	1.0	3.0
838	—	—	—	—	—	—	28.6	52.0	—	—	4.0	30.0	17.4	4.4	13.3
839	—	—	—	80.0	3,539.0	3,354.0	24.8	—	—	—	—	—	3.1	8.4	4.2
840	—	—	—	91.0	4,008.0	3,798.0	28.1	—	—	—	—	—	3.5	9.5	4.7
841	—	—	—	79.0	3,268.0	3,253.0	14.2	—	7.0	—	—	7.0	2.6	1.8	1.7
842	—	—	—	89.0	3,676.0	3,660.0	16.0	—	8.0	—	—	8.0	2.9	2.0	1.9
843	—	—	—	74.0	3,083.0	2,925.0	15.1	—	7.0	—	—	11.0	2.5	1.8	1.6
844	—	—	—	84.0	3,519.0	3,339.0	17.2	—	8.0	—	—	13.0	2.9	2.0	1.8
845	—	—	—	75.0	3,402.0	3,201.0	12.7	—	—	—	—	3.0	2.5	1.6	1.7
846	—	—	—	86.0	3,855.0	3,627.0	14.4	—	—	—	—	4.0	2.8	1.8	1.9
847	—	—	—	76.0	3,339.0	3,118.0	11.5	—	—	—	—	—	2.2	1.6	1.8
848	—	—	—	86.0	3,784.0	3,533.0	13.0	—	—	—	—	—	2.4	1.8	2.1
849	—	—	—	83.0	3,654.0	3,346.0	10.1	12.0	—	—	—	4.0	2.3	1.7	1.6
850	—	—	—	93.0	4,101.0	3,756.0	11.3	14.0	—	—	—	4.0	2.6	1.9	1.8
851	—	—	—	77.0	3,384.0	3,329.0	10.0	—	—	—	—	—	2.5	1.9	1.9
852	—	—	—	86.0	3,791.0	3,729.0	11.2	—	—	—	—	—	2.8	2.2	2.1
853	—	—	—	63.0	2,761.0	2,354.0	14.1	—	5.0	—	7.0	—	6.9	3.4	5.5
854	—	—	—	70.0	3,094.0	2,638.0	15.8	—	6.0	—	8.0	—	7.7	3.9	6.1
855	—	—	—	79.0	3,471.0	3,243.0	11.4	—	—	—	—	—	0.4	0.9	0.4
856	—	—	—	88.0	3,870.0	3,616.0	12.7	—	—	—	—	—	0.4	1.0	0.4
857	39.0	1.57	1.28	—	—	—	7.4	60.0	—	—	6.0	36.0	24.6	1.9	6.2
858	44.0	1.79	1.47	—	—	—	8.5	68.0	—	—	7.0	41.0	28.1	2.2	7.1
859	49.0	—	—	72.0	3,171.0	2,765.0	15.4	—	—	—	—	—	8.2	4.1	5.3
860	54.0	—	—	80.0	3,527.0	3,076.0	17.2	—	—	—	—	—	9.2	4.6	5.9
861	—	—	—	—	—	—	3.6	—	—	—	—	—	8.1	0.7	2.2
862	—	—	—	—	—	—	11.9	—	—	—	—	—	26.9	2.5	7.5
863	—	—	—	—	—	—	2.0	—	—	—	—	—	7.8	0.8	2.1
864	—	—	—	—	—	—	8.1	—	—	—	—	—	30.9	3.0	8.4
865	30.0	1.27	1.04	—	—	—	3.2	75.0	35.0	—	12.0	48.0	36.9	1.6	6.9
866	34.0	1.43	1.17	—	—	—	3.6	85.0	39.0	—	14.0	54.0	41.6	1.8	7.8
867	—	—	—	75.0	3,301.0	3,062.0	13.9	—	—	—	—	—	2.2	1.8	1.6
868	—	—	—	85.0	3,763.0	3,492.0	15.9	—	—	—	—	—	2.5	2.0	1.8
869	—	—	—	—	—	—	6.0	—	—	—	—	—	6.2	0.6	2.8
870	—	—	—	—	—	—	21.5	—	—	—	—	—	22.2	2.2	10.0
871	—	—	—	—	—	—	4.4	—	—	—	—	—	13.6	1.6	4.2
872	—	—	—	—	—	—	9.8	—	—	—	—	—	30.3	3.6	9.3
873	—	—	—	—	—	—	2.5	—	—	—	—	—	32.2	1.0	3.3
874	—	—	—	—	—	—	3.1	—	—	—	—	—	40.3	1.2	4.1
875	44.0	1.76	1.45	—	—	—	11.5	—	—	—	5.0	34.0	30.5	2.1	6.7
876	47.0	1.90	1.56	—	—	—	12.4	—	—	—	6.0	36.0	32.9	2.3	7.2
877	—	—	—	77.0	3,188.0	3,115.0	13.3	0.0	—	0.0	—	0.0	0.2	0.7	9.2
878	—	—	—	83.0	3,415.0	3,337.0	14.2	0.0	—	0.0	—	0.0	0.2	0.7	9.8

TABLE 1   Composition of Important Feeds: Energy Values, Proximate Analyses, Plant Cell Wall Constituents, and Acid Detergent Fiber—*Continued*

Entry Number	Feed Name Description	International Feed Number	Dry Matter (%)	Ruminants TDN (%)	DE (Mcal/kg)	ME (Mcal/kg)	NE_m (Mcal/kg)	NE_g (Mcal/kg)	Dairy Cattle NE_l (Mcal/kg)	Chickens ME_n (kcal/kg)	TME (kcal/kg)	NE_p (kcal/kg)
879	fresh (Cattle)	4-08-134	7.0	7.0	0.29	0.26	0.16	0.11	0.15	—	—	—
880			100.0	94.0	4.14	3.74	2.31	1.59	2.18	—	—	—
881	low lactose, dehy (Dried	4-01-186	93.0	74.0	3.25	2.87	1.72	1.14	1.70	2,053.0	—	1,532.0
882	whey product) (Cattle)		100.0	79.0	3.48	3.07	1.85	1.22	1.82	2,199.0	—	1,641.0
	WINTERFAT, COMMON *Eurotia lanata*											
883	fresh, stem-cured	2-26-142	80.0	28.0	1.24	0.89	0.66	—	0.59	—	—	—
884			100.0	35.0	1.54	1.11	0.83	—	0.74	—	—	—
	YEAST *Saccharomyces cerevisiae*											
885	brewers, dehy	7-05-527	93.0	74.0	3.25	2.87	1.73	1.14	1.70	2,055.0	2,943.0	1,263.0
886			100.0	79.0	3.48	3.07	1.85	1.22	1.82	2,199.0	3,150.0	1,351.0
887	irradiated, dehy	7-05-529	94.0	72.0	3.15	2.76	1.65	1.07	1.64	—	—	—
888			100.0	76.0	3.35	2.93	1.76	1.14	1.74	—	—	—
889	primary, dehy	7-05-533	93.0	71.0	3.14	2.76	1.65	1.08	1.64	—	—	—
890			100.0	77.0	3.40	2.98	1.79	1.17	1.77	—	—	—
	YEAST, TORULA *Torulopsis utilis*											
891	torula, dehy	7-05-534	93.0	73.0	3.21	2.82	1.69	1.11	1.67	1,855.0	2,872.0	—
892			100.0	78.0	3.44	3.02	1.82	1.19	1.79	1,989.0	3,080.0	—

| | Horses | | | Swine | | | | Plant Cell Wall Constituents | | | | | | | |
Entry Num- ber	TDN (%)	DE (Mcal/ kg)	ME (Mcal/ kg)	TDN (%)	DE (kcal/ kg)	ME (kcal/ kg)	Crude Pro- tein (%)	Cell Walls (%)	Cell- ulose (%)	Hemi- cell- ulose (%)	Lig- nin (%)	Acid Deter- gent Fiber (%)	Crude Fiber (%)	Ether Ex- tract (%)	Ash (%)
879	—	—	—	—	—	—	0.9	—	—	—	—	—	—	0.3	0.6
880	—	—	—	—	—	—	13.0	—	—	—	—	—	—	4.3	8.7
881	—	—	—	75.0	2,738.0	2,722.0	16.7	—	—	—	—	—	0.2	1.0	15.4
882	—	—	—	80.0	2,932.0	2,915.0	17.9	—	—	—	—	—	0.2	1.1	16.5
883	—	—	—	—	—	—	8.7	58.0	—	—	8.0	35.0	—	2.2	12.7
884	—	—	—	—	—	—	10.8	72.0	—	—	10.0	44.0	—	2.8	15.8
885	—	—	—	71.0	3,111.0	2,876.0	43.8	—	—	—	—	—	2.9	0.8	6.6
886	—	—	—	76.0	3,330.0	3,078.0	46.9	—	—	—	—	—	3.1	0.9	7.1
887	—	—	—	—	—	—	48.1	—	—	—	—	—	6.2	1.1	6.2
888	—	—	—	—	—	—	51.2	—	—	—	—	—	6.6	1.2	6.6
889	—	—	—	—	—	—	48.0	—	—	—	—	—	3.1	1.0	8.0
890	—	—	—	—	—	—	51.8	—	—	—	—	—	3.3	1.1	8.6
891	—	—	—	64.0	2,842.0	2,421.0	49.1	—	—	—	—	—	2.3	1.6	7.7
892	—	—	—	69.0	3,049.0	2,597.0	52.7	—	—	—	—	—	2.4	1.7	8.3

TABLE 2 Composition of Important Feeds: Mineral Elements, Data Expressed As-Fed and Dry (100% Dry Matter)

Entry Number	Feed Name Description	International Feed Number	Dry Matter (%)	Calcium (%)	Chlorine (%)	Magnesium (%)	Phosphorus (%)	Potassium (%)	Sodium (%)	Sulfur (%)	Cobalt (mg/kg)	Copper (mg/kg)	Iodine (mg/kg)	Iron (mg/kg)	Manganese (mg/kg)	Selenium (mg/kg)	Zinc (mg/kg)
	ALFALFA *Medicago sativa*																
001	fresh	2-00-196	24.0	0.48	0.11	0.07	0.07	0.51	0.05	0.09	0.03	2.0	—	70.0	10.0	—	4.0
002			100.0	1.96	0.47	0.27	0.30	2.09	0.19	0.37	0.13	10.0	—	286.0	43.0	—	18.0
003	hay, sun-cured	1-00-078	90.0	1.64	0.34	0.27	0.22	2.03	0.15	0.27	0.21	10.0	—	175.0	28.0	0.49	22.0
004			100.0	1.82	0.37	0.30	0.24	2.26	0.17	0.30	0.23	11.0	—	195.0	31.0	0.54	24.0
005	hay, sun-cured,	1-00-050	90.0	1.62	0.31	0.23	0.32	1.99	0.20	0.57	0.09	10.0	—	228.0	41.0	—	22.0
006	early vegetative		100.0	1.80	0.34	0.26	0.35	2.21	0.22	0.63	0.10	11.0	—	253.0	45.0	—	24.0
007	hay, sun-cured,	1-00-054	90.0	1.38	0.31	0.22	0.26	2.29	0.13	0.28	0.08	8.0	—	204.0	30.0	—	25.0
008	late vegetative		100.0	1.54	0.34	0.24	0.29	2.56	0.15	0.31	0.09	9.0	—	227.0	34.0	—	27.0
009	hay, sun-cured,	1-00-059	90.0	1.27	0.34	0.29	0.20	2.27	0.13	0.25	0.15	10.0	—	173.0	27.0	0.49	22.0
010	early bloom		100.0	1.41	0.38	0.33	0.22	2.52	0.14	0.28	0.16	11.0	—	192.0	31.0	0.54	25.0
011	hay, sun-cured,	1-00-063	90.0	1.27	0.34	0.28	0.22	1.54	0.11	0.26	0.32	13.0	—	121.0	25.0	—	21.0
012	midbloom		100.0	1.41	0.38	0.31	0.24	1.71	0.12	0.28	0.36	14.0	—	134.0	28.0	—	23.0
013	hay, sun-cured,	1-00-068	90.0	1.13	—	0.28	0.20	1.38	0.10	0.25	0.29	13.0	—	135.0	34.0	—	22.0
014	full bloom		100.0	1.25	—	0.31	0.22	1.53	0.11	0.27	0.33	14.0	—	150.0	37.0	—	25.0
015	hay, sun-cured,	1-00-071	91.0	1.03	—	0.24	0.17	1.62	0.08	0.23	0.08	13.0	—	139.0	40.0	—	22.0
016	mature		100.0	1.13	—	0.27	0.18	1.78	0.08	0.25	0.09	14.0	—	153.0	44.0	—	24.0
017	leaves, sun-cured	1-00-146	89.0	2.27	0.45	0.36	0.24	1.62	0.09	—	0.19	10.0	—	319.0	33.0	—	—
018			100.0	2.54	0.50	0.40	0.27	1.82	0.11	—	0.22	11.0	—	358.0	37.0	—	—
019	meal dehy, 15%	1-00-022	90.0	1.24	0.44	0.28	0.22	2.24	0.07	0.22	0.17	9.0	0.12	280.0	28.0	0.28	19.0
020	protein		100.0	1.37	0.48	0.31	0.24	2.48	0.08	0.24	0.19	10.0	0.13	309.0	31.0	0.31	21.0
021	meal dehy, 17%	1-00-023	92.0	1.40	0.47	0.29	0.23	2.39	0.10	0.22	0.30	10.0	0.15	405.0	31.0	0.33	19.0
022	protein		100.0	1.52	0.52	0.32	0.25	2.60	0.11	0.24	0.33	11.0	0.16	441.0	34.0	0.37	21.0
023	meal dehy, 20%	1-00-024	92.0	1.59	0.47	0.33	0.28	2.50	0.12	0.27	0.26	11.0	0.14	380.0	36.0	0.29	20.0
024	protein		100.0	1.74	0.51	0.36	0.30	2.73	0.14	0.29	0.28	12.0	0.15	415.0	39.0	0.31	22.0
025	meal dehy, 22%	1-07-851	93.0	1.69	0.52	0.31	0.30	2.40	0.12	0.30	0.31	10.0	0.17	355.0	36.0	—	19.0
026	protein		100.0	1.82	0.56	0.33	0.33	2.58	0.13	0.32	0.34	11.0	0.18	383.0	39.0	—	21.0
027	wilted silage	3-00-221	39.0	0.52	0.16	0.15	0.12	0.90	0.06	0.14	—	4.0	—	119.0	16.0	—	—
028			100.0	1.33	0.41	0.38	0.30	2.29	0.15	0.36	—	9.0	—	305.0	40.0	—	—
	ALMOND *Prunus amygdalus*																
029	hulls	4-00-359	90.0	0.21	—	—	0.10	0.47	—	0.10	—	—	—	—	—	—	—
030			100.0	0.23	—	—	0.11	0.53	—	0.11	—	—	—	—	—	—	—
	APPLES *Malus spp*																
031	pomace, oat hulls	4-28-096	89.0	0.11	—	0.06	0.10	0.43	0.12	0.02	—	—	—	266.0	7.0	—	—
032	added, dehy		100.0	0.13	—	0.07	0.12	0.49	0.14	0.02	—	—	—	299.0	8.0	—	—
	BAHIAGRASS *Paspalum notatum*																
033	fresh	2-00-464	30.0	0.14	—	0.07	0.06	0.43	—	—	—	—	—	—	—	—	—
034			100.0	0.46	—	0.25	0.22	1.45	—	—	—	—	—	—	—	—	—
035	hay, sun-cured	1-00-462	91.0	0.46	—	0.17	0.20	—	—	—	—	—	—	55.0	—	—	—
036			100.0	0.50	—	0.19	0.22	—	—	—	—	—	—	60.0	—	—	—
037	hay, sun-cured,	1-06-137	91.0	0.27	—	0.26	0.24	1.82	—	—	—	—	—	—	—	—	—
038	early vegetative (South)		100.0	0.30	—	0.29	0.26	2.00	—	—	—	—	—	—	—	—	—
039	hay, sun-cured,	1-20-787	91.0	0.26	—	0.25	0.19	1.64	—	—	—	—	—	—	—	—	—
040	late vegetative (South)		100.0	0.28	—	0.27	0.21	1.80	—	—	—	—	—	—	—	—	—
041	hay, sun-cured,	1-06-138	91.0	0.24	—	0.23	0.18	1.46	—	—	—	—	—	—	—	—	—
042	early bloom (South)		100.0	0.26	—	0.25	0.20	1.60	—	—	—	—	—	—	—	—	—

TABLE 2 Composition of Important Feeds: Mineral Elements—*Continued*

Entry Number	Feed Name Description	International Feed Number	Dry Matter (%)	Calcium (%)	Chlorine (%)	Magnesium (%)	Phosphorus (%)	Potassium (%)	Sodium (%)	Sulfur (%)	Cobalt (mg/kg)	Copper (mg/kg)	Iodine (mg/kg)	Iron (mg/kg)	Manganese (mg/kg)	Selenium (mg/kg)	Zinc (mg/kg)
	BAKERY																
043	waste, dehy	4-00-466	92.0	0.13	1.48	0.24	0.24	0.49	1.14	0.02	0.97	5.0	—	28.0	65.0	—	15.0
044	(Dried bakery product)		100.0	0.14	1.61	0.26	0.26	0.53	1.24	0.02	1.05	5.0	—	31.0	71.0	—	16.0
	BARLEY *Hordeum vulgare*																
045	grain	4-00-549	88.0	0.04	0.16	0.14	0.34	0.41	0.03	0.15	0.09	8.0	0.04	75.0	16.0	0.19	17.0
046			100.0	0.05	0.18	0.15	0.38	0.47	0.03	0.17	0.10	9.0	0.05	85.0	18.0	0.22	19.0
047	grain, Pacific	4-07-939	89.0	0.05	0.15	0.12	0.32	0.51	0.02	0.14	0.09	8.0	—	87.0	16.0	0.10	15.0
048	Coast		100.0	0.06	0.17	0.14	0.39	0.58	0.02	0.16	0.10	9.0	—	97.0	18.0	0.11	17.0
049	grain screenings	4-00-542	89.0	0.29	—	0.12	0.29	0.66	0.02	0.13	—	—	—	53.0	—	—	—
050			100.0	0.34	—	0.14	0.33	0.75	0.02	0.15	—	—	—	60.0	—	—	—
051	hay, sun-cured	1-00-495	87.0	0.20	—	0.16	0.23	1.03	0.12	0.15	0.06	21.0	—	89.0	24.0	0.14	42.0
052			100.0	0.23	—	0.18	0.26	1.18	0.14	0.17	0.07	24.0	—	101.0	27.0	0.16	48.0
053	malt sprouts,	5-00-545	94.0	0.21	0.36	0.18	0.71	0.21	1.18	0.80	—	—	—	—	32.0	0.45	—
054	dehy		100.0	0.23	0.39	0.20	0.75	0.23	1.26	0.85	—	—	—	—	34.0	0.48	—
055	straw	1-00-498	91.0	0.27	0.61	0.21	0.07	2.16	0.13	0.16	0.06	5.0	—	183.0	15.0	—	7.0
056			100.0	0.30	0.67	0.23	0.07	2.37	0.14	0.17	0.07	5.0	—	201.0	17.0	—	7.0
	BEAN, NAVY *Phaseolus vulgaris*																
057	seeds	5-00-623	89.0	0.16	0.06	0.13	0.52	1.31	0.04	0.23	—	10.0	—	99.0	21.0	—	—
058			100.0	0.18	0.06	0.15	0.59	1.47	0.05	0.26	—	11.0	—	110.0	24.0	—	—
	BEET, MANGELS *Beta vulgaris macrorhiza*																
059	roots, fresh	4-00-637	11.0	0.02	0.16	0.02	0.02	0.25	0.07	0.02	—	1.0	—	17.0	—	—	—
060			100.0	0.18	1.41	0.20	0.22	2.30	0.63	0.20	—	6.0	—	154.0	—	—	—
	BEET, SUGAR *Beta vulgaris altissima*																
061	aerial part with	3-00-660	22.0	0.35	—	0.24	0.06	1.28	0.12	0.13	—	—	—	45.0	—	—	—
062	crowns, silage molasses—see Molasses and syrup		100.0	1.56	—	1.07	0.29	5.74	0.54	0.57	—	—	—	200.0	—	—	—
063	pulp, dehy	4-00-669	91.0	0.63	0.04	0.24	0.09	0.18	0.19	0.20	0.07	12.0	—	299.0	35.0	—	1.0
064			100.0	0.69	0.04	0.27	0.10	0.20	0.21	0.22	0.08	14.0	—	329.0	38.0	—	1.0
065	pulp, wet	4-00-671	11.0	0.10	—	0.02	0.01	0.02	0.02	0.02	—	—	—	36.0	—	—	0.0
066			100.0	0.87	—	0.22	0.10	0.19	0.19	0.22	—	—	—	330.0	—	—	1.0
067	pulp with molas-	4-00-672	92.0	0.56	—	0.14	0.09	1.63	0.48	0.39	0.21	15.0	—	190.0	24.0	—	1.0
068	ses, dehy		100.0	0.61	—	0.16	0.10	1.78	0.53	0.42	0.23	16.0	—	207.0	27.0	—	2.0
	BERMUDAGRASS *Cynodon dactylon*																
069	fresh	2-00-712	34.0	0.18	—	0.06	0.07	0.57	—	—	0.03	2.0	—	—	—	—	—
070			100.0	0.53	—	0.17	0.21	1.70	—	—	0.08	6.0	—	—	—	—	—
071	hay, sun-cured	1-00-703	91.0	0.43	—	0.16	0.16	1.40	0.07	0.19	0.11	—	0.11	265.0	—	—	—
072			100.0	0.47	—	0.17	0.17	1.53	0.08	0.21	0.12	—	0.12	290.0	—	—	—
	BERMUDAGRASS, COASTAL *Cynodon dactylon*																
073	fresh	2-00-719	29.0	0.14	—	—	0.08	—	—	—	—	—	—	—	—	—	—
074			100.0	0.49	—	—	0.27	—	—	—	—	—	—	—	—	—	—

TABLE 2 Composition of Important Feeds: Mineral Elements—*Continued*

Entry Number	Feed Name Description	International Feed Number	Dry Matter (%)	Calcium (%)	Chlorine (%)	Magnesium (%)	Phosphorus (%)	Potassium (%)	Sodium (%)	Sulfur (%)	Cobalt (mg/kg)	Copper (mg/kg)	Iodine (mg/kg)	Iron (mg/kg)	Manganese (mg/kg)	Selenium (mg/kg)	Zinc (mg/kg)
075	hay, sun-cured	1-00-716	90.0	0.39	—	0.16	0.18	1.45	0.40	0.19	—	—	—	271.0	—	—	9.0
076			100.0	0.43	—	0.17	0.20	1.61	0.44	0.21	—	—	—	300.0	—	—	11.0
077	hay, sun-cured,	1-09-207	92.0	0.37	—	0.19	0.25	2.02	—	—	—	—	—	—	—	—	—
078	15 to 28 days' growth (South)		100.0	0.40	—	0.21	0.27	2.20	—	—	—	—	—	—	—	—	—
079	hay, sun-cured,	1-09-209	93.0	0.30	—	0.15	0.19	1.58	—	—	—	—	—	—	—	—	—
080	29 to 42 days' growth (South)		100.0	0.32	—	0.16	0.20	1.70	—	—	—	—	—	—	—	—	—
081	hay, sun-cured,	1-09-210	93.0	0.24	—	0.12	0.17	1.21	—	—	—	—	—	—	—	—	—
082	43 to 56 days' growth (South)		100.0	0.26	—	0.13	0.18	1.30	—	—	—	—	—	—	—	—	—
	BERMUDAGRASS, MIDLAND *Cynodon dactylon*																
083	hay, sun-cured,	1-06-139	92.0	0.31	—	0.16	0.31	2.58	—	—	—	—	—	—	—	—	—
084	15 to 28 days' growth (South)		100.0	0.34	—	0.17	0.34	2.80	—	—	—	—	—	—	—	—	—
085	hay, sun-cured,	1-06-140	92.0	0.31	—	0.13	0.24	1.93	—	—	—	—	—	—	—	—	—
086	29 to 42 days' growth (South)		100.0	0.34	—	0.14	0.26	2.10	—	—	—	—	—	—	—	—	—
	BIRDSFOOT TREFOIL— SEE TREFOIL, BIRDSFOOT																
	BLOOD																
087	meal	5-00-380	92.0	0.29	0.28	0.22	0.24	0.09	0.32	0.34	0.09	10.0	—	3,719.0	5.0	0.73	4.0
088			100.0	0.32	0.30	0.24	0.26	0.10	0.35	0.37	0.10	11.0	—	4,064.0	6.0	0.80	5.0
089	meal flash dehy	5-26-006	92.0	0.30	0.25	0.21	0.23	0.10	0.29	0.45	0.09	6.0	0.02	2,341.0	10.0	—	16.0
090			100.0	0.32	0.27	0.23	0.25	0.11	0.32	0.49	0.10	7.0	0.02	2,543.0	11.0	—	17.0
091	meal spray dehy	5-00-381	93.0	0.48	0.25	0.22	0.24	0.09	0.39	0.34	—	8.0	—	2,784.0	6.0	—	—
092	(Blood flour)		100.0	0.52	0.27	0.24	0.26	0.10	0.42	0.37	—	9.0	—	2,993.0	7.0	—	—
	BLUEGRASS, CANADA *Poa compressa*																
093	fresh	2-00-764	31.0	0.12	—	0.05	0.12	0.64	0.04	0.05	—	—	—	94.0	25.0	—	—
094			100.0	0.39	—	0.16	0.39	2.04	0.14	0.17	—	—	—	300.0	79.0	—	—
095	hay, sun-cured	1-00-762	92.0	0.28	—	0.31	0.25	1.65	0.10	0.12	—	—	—	277.0	85.0	—	—
096			100.0	0.30	—	0.33	0.27	1.78	0.11	0.13	—	—	—	300.0	93.0	—	—
	BLUEGRASS, KENTUCKY *Poa pratensis*																
097	fresh	2-00-786	35.0	0.12	0.14	0.06	0.12	0.70	0.06	0.10	—	5.0	—	90.0	19.0	—	—
098			100.0	0.33	0.40	0.17	0.34	1.98	0.16	0.29	—	14.0	—	255.0	55.0	—	—
099	hay, sun-cured	1-00-776	89.0	0.29	0.47	0.14	0.22	1.51	0.12	0.14	—	9.0	—	261.0	62.0	—	—
100			100.0	0.33	0.53	0.16	0.25	1.69	0.13	0.16	—	10.0	—	293.0	70.0	—	—
	BLUESTEM *Andropogon* spp																
101	fresh, early vegeta-	2-00-821	27.0	0.17	—	—	0.05	0.46	—	—	—	13.0	—	240.0	28.0	—	—
102	tative		100.0	0.63	—	—	0.20	1.72	—	—	—	47.0	—	895.0	106.0	—	—
103	fresh, mature	2-00-825	59.0	0.23	—	0.04	0.07	0.30	—	—	—	16.0	—	634.0	36.0	—	—
104			100.0	0.40	—	0.06	0.12	0.51	—	—	—	26.0	—	1,075.0	61.0	—	—

TABLE 2 Composition of Important Feeds: Mineral Elements—*Continued*

Entry Number	Feed Name Description	International Feed Number	Dry Matter (%)	Calcium (%)	Chlorine (%)	Magnesium (%)	Phosphorus (%)	Potassium (%)	Sodium (%)	Sulfur (%)	Cobalt (mg/kg)	Copper (mg/kg)	Iodine (mg/kg)	Iron (mg/kg)	Manganese (mg/kg)	Selenium (mg/kg)	Zinc (mg/kg)
	BREWERS																
105	grains, dehy	5-02-141	92.0	0.30	0.15	0.15	0.51	0.08	0.21	0.30	0.08	21.0	0.07	245.0	37.0	0.70	27.0
106			100.0	0.33	0.17	0.16	0.55	0.09	0.23	0.32	0.08	23.0	0.07	266.0	40.0	0.76	30.0
107	grains, wet	5-02-142	21.0	0.07	0.04	0.03	0.12	0.02	0.05	0.07	0.02	5.0	0.02	56.0	9.0	0.16	6.0
108			100.0	0.33	0.17	0.16	0.55	0.09	0.23	0.32	0.10	23.0	0.07	266.0	40.0	0.76	30.0
	BROME *Bromus spp*																
109	fresh, early	2-00-892	34.0	0.17	—	0.06	0.10	0.78	0.01	0.07	—	—	—	68.0	—	—	—
110	vegetative		100.0	0.50	—	0.18	0.30	2.30	0.02	0.20	—	—	—	200.0	—	—	—
111	fresh, mature	2-00-898	57.0	0.11	—	0.10	0.15	0.71	0.01	0.11	—	—	—	113.0	—	—	—
112			100.0	0.20	—	0.18	0.26	1.25	0.02	0.20	—	—	—	200.0	—	—	—
113	hay, sun-cured	1-00-890	91.0	0.31	—	0.09	0.17	1.74	0.02	0.18	—	—	—	181.0	—	—	—
114			100.0	0.35	—	0.09	0.19	1.93	0.02	0.20	—	—	—	200.0	—	—	—
115	hay, sun-cured,	1-00-887	88.0	0.28	—	0.08	0.33	2.04	0.02	0.18	—	—	—	—	—	—	—
116	late vegetative		100.0	0.32	—	0.09	0.37	2.32	0.02	0.20	—	—	—	—	—	—	—
117	hay, sun-cured,	1-00-888	89.0	0.27	—	0.08	0.31	2.06	0.02	0.18	—	—	—	—	—	—	—
118	late bloom		100.0	0.30	—	0.09	0.35	2.32	0.02	0.20	—	—	—	—	—	—	—
	BROME, CHEAT-GRASS *Bromus tectorum*																
119	fresh, early	2-00-908	22.0	0.14	—	—	0.06	—	—	—	—	—	—	—	—	—	—
120	vegetative		100.0	0.64	—	—	0.28	—	—	—	—	—	—	—	—	—	—
	BROME, SMOOTH *Bromus inermis*																
121	fresh, early	2-00-956	30.0	0.16	—	0.09	0.13	0.93	—	—	—	—	—	—	—	—	—
122	vegetative		100.0	0.55	—	0.32	0.45	3.16	—	—	—	—	—	—	—	—	—
123	fresh, mature	2-08-364	55.0	0.14	—	—	0.09	—	—	—	—	1.0	—	—	—	—	—
124			100.0	0.26	—	—	0.16	—	—	—	—	2.0	—	—	—	—	—
125	hay, sun-cured	1-00-947	90.0	0.25	0.32	0.17	0.20	2.06	0.02	0.17	0.08	9.0	—	110.0	54.0	0.47	26.0
126			100.0	0.28	0.35	0.18	0.22	2.28	0.02	0.19	0.09	10.0	—	122.0	59.0	0.52	29.0
	BROOMCORN MILLET—SEE MILLET, PROSO																
	BUCKWHEAT, COMMON *Fagopyrum sagittatum*																
127	grain	4-00-994	88.0	0.10	0.04	0.10	0.33	0.45	0.05	0.14	0.05	9.0	—	44.0	34.0	—	9.0
128			100.0	0.11	0.05	0.12	0.37	0.51	0.06	0.16	0.06	11.0	—	50.0	38.0	—	10.0
	BUTTERMILK																
129	condensed	5-01-159	29.0	0.42	0.13	0.15	0.30	0.26	0.26	0.03	—	0.0	—	3.0	1.0	—	13.0
130	(Cattle)		100.0	1.44	0.43	0.52	1.01	0.90	0.90	0.09	—	1.0	—	9.0	4.0	—	44.0
131	dehy (Cattle)	5-01-160	92.0	1.33	0.40	0.48	0.94	0.83	0.83	0.08	—	1.0	—	8.0	3.0	—	40.0
132			100.0	1.44	0.43	0.52	1.01	0.90	0.90	0.09	—	1.0	—	9.0	4.0	—	44.0
	CANARYGRASS, REED *Phalaris arundinacea*																
133	fresh	2-01-113	27.0	0.11	—	—	0.09	0.97	—	—	—	—	—	—	—	—	—
134			100.0	0.41	—	—	0.35	3.64	—	—	—	—	—	—	—	—	—

TABLE 2 Composition of Important Feeds: Mineral Elements—*Continued*

Entry Number	Feed Name Description	International Feed Number	Dry Matter (%)	Calcium (%)	Chlorine (%)	Magnesium (%)	Phosphorus (%)	Potassium (%)	Sodium (%)	Sulfur (%)	Cobalt (mg/kg)	Copper (mg/kg)	Iodine (mg/kg)	Iron (mg/kg)	Manganese (mg/kg)	Selenium (mg/kg)	Zinc (mg/kg)
135	hay, sun-cured	1-01-104	91.0	0.35	—	0.27	0.23	2.51	0.13	—	0.02	11.0	—	137.0	108.0	—	—
136			100.0	0.38	—	0.29	0.25	2.76	0.14	—	0.02	12.0	—	150.0	118.0	—	—
	CARROT *Daucus* spp																
137	roots, fresh	4-01-145	12.0	0.05	0.06	0.02	0.04	0.33	0.12	0.02	—	1.0	—	14.0	4.0	—	—
138			100.0	0.40	0.50	0.20	0.35	2.80	1.04	0.17	—	10.0	—	120.0	31.0	—	—
	CASEIN																
139	dehy (cattle)	5-01-162	91.0	0.61	—	0.01	0.82	0.01	0.01	—	—	4.0	—	14.0	4.0	—	27.0
140			100.0	0.67	—	0.01	0.90	0.01	0.01	—	—	4.0	—	15.0	5.0	—	30.0
	CASSAVA, COMMON *Manihot esculenta*																
141	tubers, dehy	4-09-598	88.0	0.25	—	—	0.17	0.23	—	—	—	—	—	8.0	18.0	—	—
142			100.0	0.28	—	—	0.19	0.26	—	—	—	—	—	9.0	20.0	—	—
	CATTLE *Bos taurus* buttermilk—see Buttermilk																
143	livers, fresh	5-01-166	28.0	0.01	—	0.01	0.23	0.20	0.10	—	—	6.0	—	46.0	3.0	—	27.0
144			100.0	0.04	—	0.04	0.82	0.72	0.35	—	—	22.0	—	165.0	10.0	—	95.0
145	lungs, fresh	5-07-941	21.0	0.01	—	0.01	0.15	0.07	0.15	—	0.09	1.0	0.07	69.0	0.0	0.07	12.0
146			100.0	0.06	—	0.03	0.69	0.33	0.69	—	0.42	5.0	0.31	322.0	0.0	0.35	55.0
147	manure, dehy	1-01-190	94.0	1.44	—	0.43	0.95	0.62	—	1.68	—	34.0	—	2,068.0	203.0	—	224.0
148			100.0	1.52	—	0.46	1.01	0.65	—	1.78	—	36.0	—	2,190.0	215.0	—	237.0
	milk—see Milk skim milk—see Milk																
149	spleens, fresh	5-07-942	24.0	0.01	—	0.01	0.27	0.22	0.14	—	0.24	0.0	0.18	407.0	—	—	19.0
150			100.0	0.02	—	0.05	1.13	0.91	0.58	—	1.00	1.0	0.76	1,691.0	—	—	81.0
151	udders, fresh	5-07-943	20.0	0.53	—	0.02	0.28	0.16	0.12	—	0.31	1.0	—	21.0	1.0	0.11	21.0
152			100.0	2.62	—	0.08	1.37	0.79	0.58	—	1.53	3.0	—	102.0	3.0	0.54	104.0
	CEREALS																
153	screenings	4-02-156	90.0	0.33	—	0.12	0.35	0.30	0.40	—	—	—	—	—	44.0	—	—
154			100.0	0.37	—	0.14	0.39	0.34	0.45	—	—	—	—	—	49.0	—	—
155	screenings refuse	4-02-151	91.0	0.29	—	0.22	0.34	0.18	0.25	0.30	—	—	—	245.0	—	0.65	—
156			100.0	0.32	—	0.24	0.37	0.20	0.28	0.33	—	—	—	270.0	—	0.72	—
157	screenings uncleaned	4-02-153	92.0	0.37	—	0.21	0.41	0.18	0.26	0.30	—	5.0	—	248.0	—	0.80	34.0
158			100.0	0.40	—	0.23	0.45	0.20	0.28	0.33	—	5.0	—	270.0	—	0.87	37.0
	CHICKEN *Gallus domesticus*																
159	broilers, whole, fresh	5-07-945	24.0	0.01	—	—	0.20	—	—	—	—	—	—	20.0	—	—	—
160			100.0	0.04	—	—	0.82	—	—	—	—	—	—	81.0	—	—	—
161	eggs with shells, fresh	5-01-213	43.0	9.55	—	—	0.14	—	—	—	—	—	—	—	—	—	—
162			100.0	22.20	—	—	0.33	—	—	—	—	—	—	—	—	—	—
163	feet, fresh	5-07-947	33.0	2.10	—	0.03	0.76	0.08	0.12	—	2.02	1.0	0.12	31.0	1.0	—	16.0
164			100.0	6.45	—	0.10	2.33	0.26	0.38	—	6.22	2.0	0.37	96.0	2.0	—	49.0
165	gizzards, fresh	5-07-948	25.0	0.01	—	—	0.11	0.24	0.07	—	—	—	—	29.0	—	—	—
166			100.0	0.04	—	—	0.42	0.96	0.26	—	—	—	—	116.0	—	—	—

TABLE 2 Composition of Important Feeds: Mineral Elements—*Continued*

Entry Number	Feed Name Description	International Feed Number	Dry Matter (%)	Calcium (%)	Chlorine (%)	Magnesium (%)	Phosphorus (%)	Potassium (%)	Sodium (%)	Sulfur (%)	Cobalt (mg/kg)	Copper (mg/kg)	Iodine (mg/kg)	Iron (mg/kg)	Manganese (mg/kg)	Selenium (mg/kg)	Zinc (mg/kg)
167	hens, carcass,	5-08-095	33.0	0.01	—	—	0.19	—	—	—	—	—	—	17.0	—	—	—
168	fresh		100.0	0.04	—	—	0.60	—	—	—	—	—	—	52.0	—	—	—
169	hens, whole, fresh	5-07-950	33.0	0.01	—	0.07	0.19	0.10	0.27	—	—	—	—	—	—	—	—
170			100.0	0.03	—	0.22	0.59	0.31	0.83	—	—	—	—	—	—	—	—
	CITRUS *Citrus spp*																
171	pulp fines (Dried	4-01-235	91.0	1.98	—	0.16	0.11	0.62	0.10	—	—	6.0	—	164.0	7.0	—	15.0
172	citrus meal)		100.0	2.17	—	0.18	0.12	0.68	0.11	—	—	7.0	—	180.0	8.0	—	16.0
173	pulp, silage	3-01-234	21.0	0.43	—	0.03	0.03	0.13	0.02	0.00	—	—	—	33.0	—	—	3.0
174			100.0	2.04	—	0.16	0.15	0.62	0.09	0.02	—	—	—	160.0	—	—	16.0
175	pulp without	4-01-237	91.0	1.67	—	0.16	0.11	0.72	0.08	0.08	0.14	6.0	—	345.0	7.0	—	14.0
176	fines, dehy (Dried citrus pulp) syrup—see Molasses and syrup		100.0	1.84	—	0.17	0.12	0.79	0.09	0.08	0.16	6.0	—	378.0	7.0	—	15.0
	CLOVER, ALSIKE *Trifolium hybridum*																
177	fresh	2-01-316	23.0	0.30	0.18	0.07	0.06	0.60	0.11	0.04	—	1.0	—	104.0	27.0	—	—
178			100.0	1.32	0.77	0.31	0.28	2.62	0.46	0.17	—	6.0	—	455.0	117.0	—	—
179	hay, sun-cured	1-01-313	88.0	1.13	0.69	0.36	0.23	2.17	0.40	0.17	—	5.0	—	228.0	61.0	—	—
180			100.0	1.29	0.78	0.41	0.26	2.46	0.46	0.19	—	6.0	—	260.0	69.0	—	—
	CLOVER, CRIMSON *Trifolium incarnatum*																
181	fresh	2-01-336	17.0	0.23	0.10	0.05	0.06	0.43	0.07	0.05	—	—	—	31.0	—	—	—
182			100.0	1.33	0.61	0.29	0.32	2.51	0.40	0.28	—	—	—	180.0	—	—	—
183	hay, sun-cured	1-01-328	87.0	1.22	0.55	0.24	0.19	2.09	0.34	0.24	—	—	0.06	610.0	149.0	—	—
184			100.0	1.40	0.63	0.28	0.22	2.40	0.39	0.28	—	—	0.07	700.0	171.0	—	—
	CLOVER, LADINO *Trifolium repens*																
185	fresh	2-01-383	21.0	0.27	—	0.08	0.08	0.54	0.03	0.03	—	—	—	77.0	15.0	—	—
186			100.0	1.27	—	0.37	0.35	2.56	0.12	0.12	—	—	—	361.0	72.0	—	—
187	hay, sun-cured	1-01-378	90.0	1.21	0.27	0.43	0.28	2.35	0.12	0.19	0.15	9.0	0.27	370.0	85.0	—	15.0
188			100.0	1.35	0.30	0.48	0.31	2.62	0.13	0.21	0.16	10.0	0.30	413.0	95.0	—	17.0
	CLOVER, RED *Trifolium pratense*																
189	fresh	2-01-434	23.0	0.40	0.18	0.11	0.08	0.53	0.05	0.04	0.03	2.0	—	72.0	29.0	—	—
190			100.0	1.71	0.77	0.48	0.35	2.28	0.20	0.17	0.14	9.0	—	307.0	123.0	—	—
191	fresh, regrowth	2-28-255	18.0	0.30	—	0.09	0.07	0.44	0.04	0.03	—	—	—	54.0	—	—	—
192	early vegetative		100.0	1.64	—	0.51	0.36	2.44	0.20	0.17	—	—	—	300.0	—	—	—
193	hay, sun-cured	1-01-415	89.0	1.35	0.28	0.38	0.22	1.44	0.16	0.15	0.14	10.0	0.22	163.0	65.0	—	15.0
194			100.0	1.53	0.32	0.43	0.25	1.62	0.19	0.17	0.16	11.0	0.25	184.0	73.0	—	17.0
	COCONUT *Cocos nucifera*																
195	meats, meal mech	5-01-572	92.0	0.20	—	0.31	0.61	1.50	0.04	0.34	0.13	14.0	—	1,524.0	65.0	—	—
196	extd (Copra meal)		100.0	0.22	—	0.33	0.66	1.62	0.04	0.36	0.14	15.0	—	1,651.0	71.0	—	—

TABLE 2 Composition of Important Feeds: Mineral Elements—*Continued*

Entry Number	Feed Name Description	International Feed Number	Dry Matter (%)	Calcium (%)	Chlorine (%)	Magnesium (%)	Phosphorus (%)	Potassium (%)	Sodium (%)	Sulfur (%)	Cobalt (mg/kg)	Copper (mg/kg)	Iodine (mg/kg)	Iron (mg/kg)	Manganese (mg/kg)	Selenium (mg/kg)	Zinc (mg/kg)
197	meats, meal solv	5-01-573	91.0	0.17	0.03	0.33	0.60	1.49	0.04	0.34	0.13	9.0	—	683.0	65.0	—	—
198	extd (Copra meal)		100.0	0.19	0.03	0.36	0.66	1.63	0.04	0.37	0.14	10.0	—	750.0	72.0	—	—
	COFFEE *Coffea spp*																
199	fruit with hulls	1-09-648	89.0	0.29	—	—	0.69	—	—	—	—	—	—	—	—	—	—
200	without seeds, dehy (Coffee pulp with hulls)		100.0	0.33	—	—	0.77	—	—	—	—	—	—	—	—	—	—
201	fruit without	1-09-734	87.0	0.55	—	—	0.12	2.60	0.11	—	—	5.0	—	149.0	6.0	—	4.0
202	seeds, dehy (Coffee pulp)		100.0	0.63	—	—	0.13	2.99	0.12	—	—	6.0	—	172.0	7.0	—	5.0
203	grounds, wet	1-01-576	74.0	0.09	—	—	0.06	—	—	—	—	—	—	—	—	—	—
204			100.0	0.12	—	—	0.08	—	—	—	—	—	—	—	—	—	—
	CORN *Zea mays* grain—see Corn grain, dent white, dent yellow, or flint																
	CORN, DENT YELLOW *Zea mays indentata*																
205	aerial part with	1-28-231	81.0	0.41	0.15	0.24	0.20	0.76	0.02	0.11	—	6.0	—	81.0	55.0	—	—
206	ears, sun-cured (Fodder)		100.0	0.50	0.19	0.29	0.25	0.93	0.03	0.14	—	8.0	—	100.0	68.0	—	—
207	aerial part with-	1-28-233	85.0	0.49	—	0.34	0.08	1.24	0.06	0.15	—	4.0	—	179.0	116.0	—	—
208	out ears without husks, sun-cured (Stover) (Straw)		100.0	0.57	—	0.40	0.10	1.45	0.07	0.17	—	5.0	—	210.0	136.0	—	—
209	cobs, ground	1-28-234	90.0	0.11	—	0.06	0.04	0.79	0.42	0.42	0.12	7.0	—	208.0	6.0	—	—
210			100.0	0.12	—	0.07	0.04	0.87	0.47	0.47	0.13	7.0	—	230.0	6.0	—	—
211	distillers grains,	5-28-235	94.0	0.10	0.07	0.07	0.40	0.17	0.09	0.43	0.08	45.0	0.04	209.0	22.0	0.45	33.0
212	dehy		100.0	0.11	0.08	0.07	0.43	0.18	0.10	0.46	0.09	48.0	0.05	223.0	23.0	0.48	35.0
213	distillers grains	5-28-236	92.0	0.14	0.17	0.16	0.65	0.40	0.53	0.31	0.17	53.0	—	237.0	23.0	0.39	—
214	with solubles, dehy		100.0	0.15	0.18	0.18	0.71	0.44	0.57	0.33	0.18	58.0	—	259.0	25.0	0.42	—
215	distillers solubles,	5-28-237	93.0	0.33	0.26	0.60	1.27	1.67	0.23	0.37	0.20	83.0	0.11	566.0	74.0	0.33	85.0
216	dehy		100.0	0.35	0.28	0.65	1.37	1.80	0.25	0.40	0.21	89.0	0.12	610.0	80.0	0.36	92.0
217	ears, ground	4-28-238	87.0	0.06	0.04	0.12	0.24	0.46	0.02	0.14	0.27	7.0	0.02	79.0	12.0	0.07	12.0
218	(Corn and cob meal)		100.0	0.07	0.05	0.14	0.27	0.53	0.02	0.16	0.31	8.0	0.03	91.0	14.0	0.09	14.0
219	ears with husks,	3-28-239	44.0	0.04	—	0.05	0.13	0.22	0.00	0.06	—	—	—	35.0	—	—	—
220	silage		100.0	0.10	—	0.12	0.29	0.49	0.01	0.13	—	—	—	80.0	—	—	—
221	germs, meal wet	5-28-240	91.0	0.04	0.04	0.31	0.43	0.28	0.07	0.30	—	4.0	—	337.0	4.0	0.34	104.0
222	milled solv extd		100.0	0.04	0.04	0.34	0.47	0.31	0.08	0.33	—	5.0	—	370.0	4.0	0.37	114.0
223	gluten, meal	5-28-241	91.0	0.15	0.06	0.06	0.45	0.03	0.09	0.35	0.08	28.0	—	386.0	8.0	1.01	174.0
224			100.0	0.16	0.07	0.06	0.50	0.03	0.10	0.39	0.08	30.0	—	423.0	8.0	1.11	190.0
225	gluten, meal,	5-28-242	90.0	0.07	0.09	0.08	0.48	0.19	0.06	0.65	0.05	26.0	0.02	282.0	7.0	0.83	31.0
226	60% protein		100.0	0.08	0.10	0.09	0.54	0.21	0.06	0.72	0.05	29.0	0.02	313.0	7.0	0.92	35.0
227	gluten with bran	5-28-243	90.0	0.33	0.22	0.33	0.74	0.57	0.94	0.21	0.09	47.0	0.07	424.0	23.0	0.27	65.0
228	(Corn gluten feed)		100.0	0.36	0.25	0.36	0.82	0.64	1.05	0.23	0.10	52.0	0.07	471.0	26.0	0.30	72.0
229	grain	4-02-935	89.0	0.03	0.04	0.12	0.26	0.33	0.03	0.11	0.05	4.0	—	27.0	5.0	0.07	13.0
230			100.0	0.03	0.05	0.14	0.29	0.37	0.03	0.12	0.05	4.0	—	30.0	5.0	0.08	14.0

TABLE 2 Composition of Important Feeds: Mineral Elements—*Continued*

Entry Number	Feed Name Description	International Feed Number	Dry Matter (%)	Calcium (%)	Chlorine (%)	Magnesium (%)	Phosphorus (%)	Potassium (%)	Sodium (%)	Sulfur (%)	Cobalt (mg/kg)	Copper (mg/kg)	Iodine (mg/kg)	Iron (mg/kg)	Manganese (mg/kg)	Selenium (mg/kg)	Zinc (mg/kg)
231	grain, high	4-20-770	77.0	0.01	0.04	0.11	0.25	0.27	0.01	0.11	—	3.0	—	23.0	5.0	—	14.0
232	moisture		100.0	0.02	0.05	0.14	0.32	0.35	0.01	0.14	—	4.0	—	30.0	6.0	—	18.0
233	grain, opaque 2	4-28-253	90.0	0.03	—	0.13	0.20	0.35	—	0.10	—	—	—	—	—	—	—
234	(High lysine)		100.0	0.03	—	0.14	0.22	0.39	—	0.11	—	—	—	—	—	—	—
235	grits (Hominy	4-03-010	88.0	0.01	0.05	0.02	0.09	0.08	0.01	0.17	—	—	—	15.0	—	—	—
236	grits)		100.0	0.01	0.06	0.02	0.10	0.09	0.01	0.19	—	—	—	17.0	—	—	—
237	grits by-product	4-03-011	90.0	0.05	0.05	0.24	0.52	0.59	0.08	0.03	0.05	14.0	—	67.0	15.0	0.10	3.0
238	(Hominy feed)		100.0	0.05	0.06	0.26	0.57	0.65	0.09	0.03	0.06	15.0	—	75.0	16.0	0.11	3.0
239	silage	3-02-912	30.0	0.09	0.05	0.06	0.08	0.36	0.01	0.04	0.03	3.0	—	48.0	12.0	—	6.0
240			100.0	0.29	0.18	0.21	0.26	1.18	0.02	0.13	0.10	9.0	—	160.0	39.0	—	21.0
241	silage, aerial part	3-28-251	31.0	0.12	—	0.10	0.09	0.47	0.01	0.03	—	—	—	—	—	—	—
242	without ears without husks (Stalklage) (Stover)		100.0	0.38	—	0.31	0.31	1.54	0.03	0.11	—	—	—	—	—	—	—
243	silage, few ears	3-28-245	29.0	0.10	—	0.07	0.06	0.41	—	0.02	—	—	—	—	—	—	—
244			100.0	0.34	—	0.23	0.19	1.41	—	0.08	—	—	—	—	—	—	—
245	silage, well eared	3-28-250	33.0	0.08	—	0.06	0.07	0.32	0.00	0.05	0.02	3.0	—	87.0	10.0	—	7.0
246			100.0	0.23	—	0.19	0.22	0.96	0.01	0.15	0.06	10.0	—	260.0	30.0	—	21.0
247	silage, milk stage	3-08-402	22.0	0.09	—	0.09	0.07	0.35	0.00	—	—	—	—	—	—	—	—
248			100.0	0.41	—	0.41	0.29	1.57	0.01	—	—	—	—	—	—	—	—
249	silage, dough	3-28-246	26.0	0.07	—	0.05	0.05	0.25	0.00	0.04	—	—	—	48.0	—	—	—
250	stage		100.0	0.27	—	0.18	0.19	0.95	0.01	0.14	—	—	—	180.0	—	—	—
	CORN, DENT WHITE *Zea mays indentata*																
251	grits, by-product	4-02-990	90.0	0.03	—	0.23	0.69	0.64	0.07	0.03	0.02	13.0	—	71.0	14.0	—	—
252	(Hominy feed)		100.0	0.04	—	0.26	0.77	0.71	0.08	0.03	0.02	15.0	—	79.0	15.0	—	—
	CORN, FLINT *Zea mays indurata*																
253	grain	4-02-948	89.0	—	—	—	0.27	0.32	—	—	—	12.0	—	27.0	7.0	—	—
254			100.0	—	—	—	0.31	0.36	—	—	—	13.0	—	30.0	8.0	—	—
	CORN, SWEET *Zea mays saccharata*																
255	process residue,	2-02-975	77.0	0.23	—	0.18	0.55	0.88	0.02	0.10	—	5.0	—	154.0	—	—	—
256	fresh		100.0	0.30	—	0.24	0.72	1.15	0.03	0.13	—	7.0	—	200.0	—	—	—
257	process residue,	3-07-955	32.0	0.10	—	0.08	0.29	0.36	0.01	0.04	—	—	—	63.0	—	—	—
258	silage		100.0	0.30	—	0.24	0.90	1.15	0.03	0.11	—	—	—	200.0	—	—	—
	COTTON *Gossypium* spp																
259	bolls, sun-cured	1-01-596	92.0	0.83	—	0.26	0.11	2.50	—	—	—	—	—	—	—	—	—
260			100.0	0.90	—	0.28	0.12	2.73	—	—	—	—	—	—	—	—	—
261	hulls	1-01-599	91.0	0.13	0.02	0.13	0.09	0.79	0.02	0.08	0.02	12.0	—	119.0	108.0	—	20.0
262			100.0	0.15	0.02	0.14	0.09	0.87	0.02	0.09	0.02	13.0	—	131.0	119.0	—	22.0
263	seeds, with lint	5-01-614	92.0	0.14	—	0.32	0.68	1.11	0.29	0.24	—	50.0	—	138.0	10.0	—	—
264			100.0	0.16	—	0.35	0.75	1.21	0.31	0.26	—	54.0	—	151.0	10.0	—	—
265	seeds, meal mech	5-01-609	93.0	0.20	0.02	0.53	0.90	1.26	0.04	0.24	0.15	21.0	—	139.0	22.0	—	—
266	extd (Whole pressed cotton-seed)		100.0	0.21	0.02	0.57	0.97	1.35	0.04	0.26	0.16	23.0	—	150.0	24.0	—	—

TABLE 2 Composition of Important Feeds: Mineral Elements—*Continued*

Entry Number	Feed Name Description	International Feed Number	Dry Matter (%)	Calcium (%)	Chlorine (%)	Magnesium (%)	Phosphorus (%)	Potassium (%)	Sodium (%)	Sulfur (%)	Cobalt (mg/kg)	Copper (mg/kg)	Iodine (mg/kg)	Iron (mg/kg)	Manganese (mg/kg)	Selenium (mg/kg)	Zinc (mg/kg)
267	seeds, meal mech	5-01-625	92.0	0.18	—	0.53	0.96	1.35	0.04	0.26	0.15	19.0	—	181.0	23.0	—	—
268	extd, 36% protein		100.0	0.20	—	0.58	1.04	1.46	0.05	0.28	0.16	20.0	—	197.0	25.0	—	—
269	seeds, meal mech	5-01-617	93.0	0.19	0.04	0.54	1.08	1.34	0.04	0.40	0.16	19.0	—	182.0	22.0	—	64.0
270	extd, 41% protein		100.0	0.21	0.05	0.58	1.16	1.45	0.05	0.43	0.17	20.0	—	197.0	24.0	—	69.0
271	seeds, meal pre-	5-07-872	91.0	0.20	0.04	0.50	1.09	1.26	0.04	0.31	0.74	18.0	—	202.0	20.0	—	63.0
272	pressed solv extd, 41% protein		100.0	0.22	0.04	0.55	1.21	1.39	0.04	0.34	0.82	20.0	—	223.0	23.0	—	69.0
273	seeds, meal pre-	5-07-873	91.0	0.15	—	—	0.91	—	—	—	—	—	—	—	—	—	—
274	pressed solv extd, 44% protein		100.0	0.17	—	—	1.00	—	—	—	—	—	—	—	—	—	—
275	seeds, meal solv	5-01-621	91.0	0.17	0.05	0.54	1.10	1.39	0.04	0.26	0.15	20.0	—	208.0	21.0	—	61.0
276	extd, 41% protein		100.0	0.18	0.05	0.59	1.21	1.52	0.05	0.28	0.17	22.0	—	228.0	23.0	—	68.0
277	seeds without	5-07-874	93.0	0.18	0.05	0.46	1.16	1.45	0.05	0.52	0.04	15.0	—	112.0	23.0	—	74.0
278	hulls, meal prepressed solv extd, 50% protein		100.0	0.19	0.05	0.50	1.24	1.56	0.06	0.56	0.05	16.0	—	120.0	25.0	—	79.0
	COWPEA, COMMON *Vigna sinensis*																
279	hay, sun-cured	1-01-645	90.0	1.26	0.15	0.41	0.31	2.03	0.24	0.32	0.06	—	—	270.0	—	—	—
280			100.0	1.40	0.17	0.45	0.35	2.26	0.27	0.35	0.07	—	—	300.0	—	—	—
	CRAB *Callinectes sapidus–Cancer* spp																
281	process residue,	5-01-663	92.0	14.56	1.51	0.94	1.59	0.45	0.88	0.25	—	33.0	0.56	4,356.0	133.0	—	—
282	meal (Crab meal)		100.0	15.77	1.63	1.02	1.72	0.49	0.95	0.27	—	35.0	0.60	4,719.0	144.0	—	—
	DISTILLERS GRAINS—SEE CORN, SEE SORGHUM																
	DROPSEED, SAND *Sporobolus cryptandrus*																
283	fresh, stem-cured	2-05-596	88.0	0.51	—	—	0.05	—	—	—	—	—	—	—	—	—	—
284			100.0	0.57	—	—	0.06	—	—	—	—	—	—	—	—	—	—
	EMMER *Triticum dicoccum*																
285	grain	4-01-830	91.0	0.05	—	—	0.36	0.47	—	—	—	31.0	—	54.0	78.0	—	—
286			100.0	0.06	—	—	0.40	0.52	—	—	—	35.0	—	60.0	86.0	—	—
	FESCUE *Festuca* spp																
287	hay, sun-cured,	1-06-132	91.0	0.46	—	0.20	0.33	2.09	—	—	—	—	—	—	—	—	—
288	early vegetative (South)		100.0	0.51	—	0.22	0.36	2.30	—	—	—	—	—	—	—	—	—

TABLE 2   Composition of Important Feeds: Mineral Elements—*Continued*

Entry Number	Feed Name Description	International Feed Number	Dry Matter (%)	Calcium (%)	Chlorine (%)	Magnesium (%)	Phosphorus (%)	Potassium (%)	Sodium (%)	Sulfur (%)	Cobalt (mg/kg)	Copper (mg/kg)	Iodine (mg/kg)	Iron (mg/kg)	Manganese (mg/kg)	Selenium (mg/kg)	Zinc (mg/kg)
289	hay, sun-cured	1-13-582	91.0	0.36	—	0.18	0.31	1.82	—	—	—	—	—	—	—	—	—
290	late vegetative (South)		100.0	0.40	—	0.20	0.34	2.00	—	—	—	—	—	—	—	—	—
291	hay, sun-cured	1-01-871	92.0	0.28	—	0.18	0.24	1.56	—	—	—	—	—	—	—	—	—
292	early bloom (South)		100.0	0.30	—	0.19	0.26	1.70	—	—	—	—	—	—	—	—	—
	FESCUE, ALTA *Festuca arundinacea*																
293	hay, sun-cured	1-05-684	91.0	0.36	—	0.28	0.20	2.23	0.06	—	—	—	—	—	—	—	—
294			100.0	0.40	—	0.31	0.22	2.45	0.06	—	—	—	—	—	—	—	—
	FESCUE, KENTUCKY 31 *Festuca arundinacea*																
295	fresh, vegetative	2-01-902	29.0	0.15	—	—	0.11	—	—	—	—	—	—	—	—	—	—
296			100.0	0.51	—	—	0.37	—	—	—	—	—	—	—	—	—	—
	FESCUE, MEADOW *Festuca elatior*																
297	fresh	2-01-920	28.0	0.17	—	0.10	0.12	0.65	—	—	0.04	1.0	—	—	—	—	—
298			100.0	0.61	—	0.37	0.42	2.34	—	—	0.13	4.0	—	—	—	—	—
299	hay, sun-cured	1-01-912	88.0	0.35	—	0.44	0.27	1.61	—	—	0.12	—	—	—	22.0	—	—
300			100.0	0.40	—	0.50	0.31	1.84	—	—	0.14	—	—	—	25.0	—	—
	FISH																
301	solubles, condensed	5-01-969	50.0	0.22	2.70	0.03	0.59	1.61	2.34	0.12	0.07	46.0	1.11	223.0	13.0	1.97	44.0
302			100.0	0.43	5.38	0.06	1.18	3.22	4.67	0.25	0.14	92.0	2.21	445.0	27.0	3.92	87.0
303	solubles, dehy	5-01-971	93.0	1.29	—	0.30	1.49	0.37	0.37	—	—	—	—	302.0	50.0	—	77.0
304			100.0	1.39	—	0.32	1.60	0.40	0.40	—	—	—	—	326.0	54.0	—	83.0
	FISH, ALEWIFE *Pomolobus pseudoharengus*																
305	meal mech extd	5-09-830	90.0	5.95	—	0.17	3.18	0.66	0.26	—	—	21.0	—	680.0	22.0	—	110.0
306			100.0	6.63	—	0.18	3.54	0.73	0.29	—	—	23.0	—	756.0	24.0	—	122.0
	FISH, ANCHOVY *Engraulis ringen*																
307	meal mech extd	5-01-985	92.0	3.75	1.00	0.25	2.49	0.72	0.88	0.77	0.17	9.0	3.13	218.0	11.0	1.35	105.0
308			100.0	4.08	1.08	0.27	2.70	0.78	0.95	0.84	0.19	10.0	3.41	237.0	12.0	1.47	114.0
	FISH, CARP *Cyprinus carpio*																
309	whole, fresh	5-01-986	31.0	0.07	—	—	0.35	0.40	0.07	—	—	—	—	12.0	—	—	—
310			100.0	0.23	—	—	1.14	1.29	0.23	—	—	—	—	40.0	—	—	—
	FISH, CATFISH *Ictalurus* spp																
311	cuttings, fresh	5-09-832	34.0	1.87	—	—	—	—	—	—	—	2.0	—	—	—	—	—
312			100.0	5.57	—	—	—	—	—	—	—	7.0	—	—	—	—	—
313	meal mech extd	5-09-835	92.0	7.18	—	—	—	—	—	—	—	26.0	—	—	—	—	—
314			100.0	7.77	—	—	—	—	—	—	—	28.0	—	—	—	—	—

TABLE 2 Composition of Important Feeds: Mineral Elements—*Continued*

Entry Number	Feed Name Description	International Feed Number	Dry Matter (%)	Calcium (%)	Chlorine (%)	Magnesium (%)	Phosphorus (%)	Potassium (%)	Sodium (%)	Sulfur (%)	Cobalt (mg/kg)	Copper (mg/kg)	Iodine (mg/kg)	Iron (mg/kg)	Manganese (mg/kg)	Selenium (mg/kg)	Zinc (mg/kg)
	FISH, HAKE *Merluccius* spp–*Urophycis* spp																
315	whole, fresh	5-07-969	20.0	0.61	—	—	0.39	—	—	—	—	—	—	—	—	—	—
316			100.0	3.06	—	—	1.93	—	—	—	—	—	—	—	—	—	—
	FISH, HERRING *Clupea harengus*																
317	meal mech extd	5-02-000	92.0	2.20	0.99	0.15	1.68	1.08	0.60	0.46	0.05	6.0	0.53	125.0	6.0	1.90	131.0
318			100.0	2.40	1.08	0.16	1.82	1.17	0.66	0.50	0.06	6.0	0.57	136.0	6.0	2.07	143.0
319	whole, fresh	5-01-999	26.0	—	—	—	0.25	0.53	0.09	—	—	—	—	13.0	—	—	—
320			100.0	—	—	—	0.96	2.04	0.36	—	—	—	—	50.0	—	—	—
	FISH, MACKEREL, ATLANTIC *Scomber scombrus*																
321	whole, fresh	5-07-971	30.0	1.10	—	0.03	0.39	0.17	0.17	—	0.09	1.0	0.23	27.0	—	—	24.0
322			100.0	3.64	—	0.10	1.28	0.55	0.56	—	0.29	3.0	0.76	90.0	—	—	78.0
	FISH, MENHADEN *Brevoortia tyrannus*																
323	meal mech extd	5-02-009	92.0	5.18	0.55	0.14	2.89	0.70	0.39	0.45	0.15	11.0	1.09	480.0	34.0	2.19	148.0
324			100.0	5.65	0.60	0.16	3.16	0.76	0.43	0.49	0.17	12.0	1.19	524.0	37.0	2.40	162.0
	FISH, REDFISH *Sciaenops ocellata*																
325	meal mech extd	5-07-973	93.0	6.48	—	—	3.39	—	—	—	—	—	—	—	8.0	1.77	—
326			100.0	6.96	—	—	3.64	—	—	—	—	—	—	—	8.0	1.90	—
327	whole, fresh	5-08-191	24.0	—	—	—	—	0.30	0.07	—	—	—	—	—	—	—	—
328			100.0	—	—	—	—	1.26	0.30	—	—	—	—	—	—	—	—
	FISH, SALMON *Oncorhynchus* spp–*Salmo* spp																
329	meal mech extd	5-02-012	93.0	5.47	—	—	3.46	—	—	—	0.06	12.0	—	179.0	8.0	1.78	—
330			100.0	5.88	—	—	3.72	—	—	—	0.07	13.0	—	193.0	8.0	1.91	—
331	whole, fresh	5-02-011	36.0	0.08	—	—	0.25	0.40	0.04	—	—	—	—	10.0	—	—	—
332			100.0	0.22	—	—	0.68	1.12	0.11	—	—	—	—	27.0	—	—	—
	FISH, SARDINE *Clupea* spp–*Sardinops* spp																
333	meal mech extd	5-02-015	93.0	4.61	0.41	0.10	2.68	0.32	0.18	0.31	0.18	20.0	—	299.0	23.0	1.77	—
334			100.0	4.95	0.44	0.11	2.88	0.35	0.19	0.33	0.20	22.0	—	321.0	25.0	1.90	—
	FISH, SOLE *Soleidae* (family)																
335	whole, fresh	5-07-976	20.0	0.63	—	—	0.39	—	—	—	—	—	—	—	—	—	—
336			100.0	3.19	—	—	2.00	—	—	—	—	—	—	—	—	—	—
	FISH, TUNA *Thunnus thynnus*–*Thunnus albacares*																
337	meal mech extd	5-02-023	93.0	7.86	1.01	0.23	4.21	0.72	0.74	0.68	0.18	10.0	—	355.0	8.0	4.30	211.0
338			100.0	8.48	1.09	0.25	4.54	0.77	0.80	0.73	0.19	11.0	—	383.0	9.0	4.64	227.0

TABLE 2 Composition of Important Feeds: Mineral Elements—*Continued*

Entry Number	Feed Name Description	International Feed Number	Dry Matter (%)	Calcium (%)	Chlorine (%)	Magnesium (%)	Phosphorus (%)	Potassium (%)	Sodium (%)	Sulfur (%)	Cobalt (mg/kg)	Copper (mg/kg)	Iodine (mg/kg)	Iron (mg/kg)	Manganese (mg/kg)	Selenium (mg/kg)	Zinc (mg/kg)
	FISH, TURBOT *Psetta maxima*																	
339	whole, fresh	5-07-978	25.0	0.31	—	—	0.22	—	—	—	—	—	—	—	—	—	—
340			100.0	1.24	—	—	0.88	—	—	—	—	—	—	—	—	—	—
	FISH, WHITE *Gadidae* (family)– *Lophiidae* (family)																	
341	meal mech extd	5-02-025	91.0	7.31	0.50	0.18	3.81	0.83	0.78	0.48	—	6.0	—	181.0	12.0	1.62	90.0
342			100.0	8.02	0.55	0.20	4.17	0.91	0.85	0.53	—	6.0	—	199.0	14.0	1.77	98.0
	FLAX *Linum usitatissimum*																	
342	seed screenings	4-02-056	91.0	0.34	—	0.39	0.43	0.77	—	0.23	—	—	—	91.0	—	—	—
344			100.0	0.37	—	0.43	0.47	0.84	—	0.25	—	—	—	100.0	—	—	—
345	seeds, meal mech extd (Linseed meal)	5-02-045	91.0	0.41	0.04	0.58	0.87	1.22	0.11	0.37	0.41	26.0	0.07	176.0	38.0	0.81	33.0
346			100.0	0.45	0.04	0.64	0.96	1.34	0.12	0.41	0.46	29.0	0.07	194.0	42.0	0.89	36.0
347	seeds, meal solv extd (Linseed meal)	5-02-048	90.0	0.39	0.04	0.60	0.80	1.38	0.14	0.39	0.19	26.0	—	319.0	38.0	0.82	—
348			100.0	0.43	0.04	0.66	0.89	1.53	0.15	0.43	0.21	29.0	—	354.0	42.0	0.91	—
	GALLETA *Hilaria jamesii*																	
349	fresh, stem-cured	2-05-594	71.0	0.74	—	0.07	0.05	0.71	0.01	0.07	—	—	—	71.0	—	—	—
350			100.0	1.05	—	0.10	0.07	1.00	0.01	0.10	—	—	—	100.0	—	—	—
	GELATIN																	
351	process residue (Gelatin by-products)	5-14-503	90.0	0.49	—	0.05	—	—	—	—	—	—	—	—	—	—	—
352			100.0	0.55	—	0.05	—	—	—	—	—	—	—	—	—	—	—
	GRAMA *Bouteloua* spp																	
353	fresh, early vegetative	2-02-163	41.0	0.22	—	—	0.08	—	—	—	—	2.0	—	—	18.0	—	—
354			100.0	0.53	—	—	0.19	—	—	—	—	6.0	—	—	44.0	—	—
355	fresh, mature	2-02-166	63.0	0.22	—	—	0.08	0.22	—	—	0.12	8.0	—	824.0	30.0	—	—
356			100.0	0.34	—	—	0.12	0.35	—	—	0.18	13.0	—	1,300.0	47.0	—	—
	GRAPE *Vitis* spp																	
357	marc, dehy (Pomace)	1-02-208	91.0	0.55	0.01	—	0.05	0.56	0.08	—	—	—	—	—	37.0	—	22.0
358			100.0	0.61	0.01	—	0.06	0.62	0.09	—	—	—	—	—	41.0	—	24.0
	GROUNDNUT— SEE PEANUT																	
	HEMICELLULOSE EXTRACT (MASONEX)																	
359		4-08-030	76.0	0.79	—	—	0.07	—	—	—	—	—	—	—	—	—	—
360			100.0	1.03	—	—	0.09	—	—	—	—	—	—	—	—	—	—
	HOG MILLET— SEE MILLET, PROSO																	

TABLE 2 Composition of Important Feeds: Mineral Elements—*Continued*

Entry Number	Feed Name Description	International Feed Number	Dry Matter (%)	Calcium (%)	Chlorine (%)	Magnesium (%)	Phosphorus (%)	Potassium (%)	Sodium (%)	Sulfur (%)	Cobalt (mg/kg)	Copper (mg/kg)	Iodine (mg/kg)	Iron (mg/kg)	Manganese (mg/kg)	Selenium (mg/kg)	Zinc (mg/kg)
	HOMINY FEED—SEE CORN																
	HORSE *Equus caballus*																
361	meat, fresh	5-07-980	29.0	0.02	—	0.01	0.31	0.11	0.05	—	0.06	—	0.09	49.0	0.0	—	18.0
362			100.0	0.07	—	0.04	1.06	0.38	0.18	—	0.21	—	0.29	167.0	0.0	—	60.0
	JOHNSONGRASS— SEE SORGHUM, JOHNSONGRASS																
	KENTUCKY BLUEGRASS— SEE BLUEGRASS, KENTUCKY																
	LESPEDEZA, COMMON *Lespedeza striata*																
363	hay, sun-cured,	1-02-554	92.0	1.08	—	0.24	0.22	0.92	—	—	—	—	—	284.0	204.0	—	—
364	midbloom		100.0	1.18	—	0.26	0.24	1.01	—	—	—	—	—	310.0	223.0	—	—
365	hay, sun-cured,	1-20-887	89.0	1.02	—	0.20	0.19	0.93	—	—	—	—	—	268.0	102.0	—	36.0
366	full bloom		100.0	1.14	—	0.23	0.21	1.04	—	—	—	—	—	300.0	114.0	—	41.0
	LESPEDEZA, COMMON– LESPEDEZA, KOREAN *Lespedeza striata–Lespedeza stipulacea*																
367	fresh	2-26-029	30.0	0.41	—	0.08	0.06	0.34	—	—	—	—	—	75.0	—	—	—
368			100.0	1.35	—	0.27	0.21	1.12	—	—	—	—	—	250.0	—	—	—
369	hay, sun-cured	1-26-035	93.0	1.14	—	0.23	0.23	0.93	—	—	0.04	0.0	—	297.0	186.0	—	—
370			100.0	1.23	—	0.25	0.25	1.00	—	—	0.04	0.0	—	320.0	200.0	—	—
	LESPEDEZA, KOREAN *Lespedeza stipulacea*																
371	fresh	2-02-598	30.0	0.37	—	—	0.12	0.49	—	—	—	—	—	—	—	—	—
372			100.0	1.23	—	—	0.39	1.62	—	—	—	—	—	—	—	—	—
373	hay, sun-cured	1-02-592	91.0	1.05	—	0.25	0.25	0.94	—	—	—	—	—	312.0	107.0	—	—
374			100.0	1.15	—	0.27	0.27	1.03	—	—	—	—	—	343.0	118.0	—	—
	LESPEDEZA, CHINESE *Lespedeza cuneata*																
375	fresh	2-02-611	35.0	0.44	—	0.08	0.10	0.41	—	—	0.02	—	—	85.0	36.0	—	—
376			100.0	1.28	—	0.22	0.28	1.19	—	—	0.07	—	—	245.0	104.0	—	—
377	hay, sun-cured	1-02-607	92.0	1.09	—	0.21	0.22	0.85	—	—	—	—	—	266.0	100.0	—	—
378			100.0	1.19	—	0.23	0.24	0.93	—	—	—	—	—	290.0	109.0	—	—
	LIGNIN SULFONATE, CALCIUM																
379	dehy	8-16-028	97.0	3.60	—	—	—	—	—	4.40	—	—	—	—	—	—	—
380			100.0	3.75	—	—	—	—	—	4.50	—	—	—	—	—	—	—
	LINSEED—SEE FLAX																
	LIVERS																
381	meal	5-00-389	92.0	0.56	—	—	1.26	—	—	—	0.13	89.0	—	630.0	9.0	—	—
382			100.0	0.61	—	—	1.36	—	—	—	0.15	97.0	—	681.0	10.0	—	—

TABLE 2   Composition of Important Feeds: Mineral Elements—*Continued*

Entry Number	Feed Name Description	International Feed Number	Dry Matter (%)	Calcium (%)	Chlorine (%)	Magnesium (%)	Phosphorus (%)	Potassium (%)	Sodium (%)	Sulfur (%)	Cobalt (mg/kg)	Copper (mg/kg)	Iodine (mg/kg)	Iron (mg/kg)	Manganese (mg/kg)	Selenium (mg/kg)	Zinc (mg/kg)
	MAIZE—SEE CORN																
	MANGELS—SEE BEET																
	MANURE—SEE CATTLE, SEE POULTRY																
	MASONEX—SEE HEMICELLULOSE EXTRACT																
	MEADOW PLANTS, INTERMOUNTAIN																
383	hay, sun-cured	1-03-181	95.0	0.58	—	0.16	0.17	1.50	0.11	—	—	—	—	—	—	—	—
384			100.0	0.61	—	0.17	0.18	1.58	0.12	—	—	—	—	—	—	—	—
	MEAT																
385	meal rendered	5-00-385	94.0	8.85	1.19	0.27	4.44	0.57	1.29	0.47	0.13	10.0	—	440.0	10.0	0.44	80.0
386			100.0	9.44	1.27	0.29	4.74	0.61	1.37	0.50	0.14	10.0	—	470.0	10.0	0.47	85.0
387	with blood, meal rendered (Tankage)	5-00-386	92.0	5.86	1.73	0.36	3.07	0.55	1.67	0.70	0.15	39.0	—	2,103.0	19.0	—	—
388			100.0	6.37	1.88	0.39	3.33	0.60	1.81	0.76	0.17	42.0	—	2,283.0	21.0	—	—
389	with blood with bone, meal rendered (Tankage)	5-00-387	93.0	11.16	—	—	5.41	—	—	0.26	—	—	—	—	—	0.26	—
390			100.0	12.01	—	—	5.82	—	—	0.28	—	—	—	—	—	0.28	—
391	with bone, meal rendered	5-00-388	93.0	10.30	0.74	1.02	5.10	1.33	0.72	0.25	0.18	2.0	1.31	684.0	13.0	0.26	89.0
392			100.0	11.06	0.80	1.09	5.48	1.43	0.77	0.27	0.19	2.0	1.41	735.0	14.0	0.28	96.0
	MILK																
393	dehy (Cattle)	5-01-167	96.0	0.91	0.88	0.09	0.71	1.04	0.37	0.31	0.01	1.0	—	10.0	0.0	—	22.0
394			100.0	0.95	0.92	0.10	0.74	1.08	0.38	0.32	0.01	1.0	—	10.0	0.0	—	23.0
395	fresh (Cattle)	5-01-168	12.0	0.12	0.11	0.01	0.09	0.14	0.05	0.04	0.00	0.0	—	1.0	—	—	3.0
396			100.0	0.95	0.92	0.10	0.76	1.12	0.38	0.32	0.01	1.0	—	10.0	—	—	23.0
397	skimmed dehy (Cattle)	5-01-175	94.0	1.28	0.90	0.12	1.02	1.59	0.46	0.32	0.11	1.0	—	9.0	2.0	0.13	38.0
398			100.0	1.36	0.96	0.13	1.09	1.70	0.49	0.34	0.12	1.0	—	10.0	2.0	0.13	41.0
399	skimmed fresh (Cattle)	5-01-170	10.0	0.13	0.09	0.01	0.10	0.18	0.04	0.03	0.01	—	—	1.0	0.0	—	5.0
400			100.0	1.31	0.96	0.12	1.04	1.90	0.47	0.32	0.11	—	—	10.0	2.0	—	51.0
	MILLET, FOX-TAIL *Setaria italica*																
401	fresh	2-03-101	28.0	0.09	—	—	0.05	0.55	—	—	—	—	—	—	—	—	—
402			100.0	0.32	—	—	0.19	1.94	—	—	—	—	—	—	—	—	—
403	grain	4-03-102	89.0	—	—	—	0.20	0.31	—	—	—	—	—	—	—	—	—
404			100.0	—	—	—	0.22	0.35	—	—	—	—	—	—	—	—	—
405	hay, sun-cured	1-03-099	87.0	0.29	0.11	0.20	0.17	1.69	0.09	0.14	—	—	—	—	119.0	—	—
406			100.0	0.33	0.13	0.23	0.19	1.94	0.10	0.16	—	—	—	—	136.0	—	—
	MILLET, PROSO *Panicum miliaceum*																
407	grain	4-03-120	90.0	0.03	—	0.16	0.30	0.43	—	—	—	—	—	71.0	—	—	—
408			100.0	0.03	—	0.18	0.34	0.48	—	—	—	—	—	79.0	—	—	—
	MOLASSES AND SYRUP																
409	beet, sugar, molasses, more than 48% invert sugar more than 79.5 degrees brix	4-00-668	78.0	0.13	1.28	0.23	0.03	4.72	1.15	0.46	0.36	17.0	—	68.0	4.0	—	14.0
410			100.0	0.17	1.64	0.29	0.03	6.07	1.48	0.60	0.46	22.0	—	87.0	6.0	—	18.0

TABLE 2 Composition of Important Feeds: Mineral Elements—*Continued*

Entry Number	Feed Name Description	International Feed Number	Dry Matter (%)	Calcium (%)	Chlorine (%)	Magnesium (%)	Phosphorus (%)	Potassium (%)	Sodium (%)	Sulfur (%)	Cobalt (mg/kg)	Copper (mg/kg)	Iodine (mg/kg)	Iron (mg/kg)	Manganese (mg/kg)	Selenium (mg/kg)	Zinc (mg/kg)
411	citrus, syrup (Citrus molasses)	4-01-241	68.0	1.16	0.07	0.14	0.09	0.09	0.28	0.16	0.11	73.0	—	344.0	26.0	—	93.0
412			100.0	1.72	0.11	0.21	0.13	0.14	0.41	0.23	0.16	108.0	—	508.0	38.0	—	137.0
413	sugarcane, molasses, dehy	4-04-695	94.0	1.04	—	0.44	0.14	3.40	0.19	0.43	1.15	75.0	1.98	236.0	54.0	—	31.0
414	lasses, dehy		100.0	1.10	—	0.47	0.15	3.60	0.20	0.46	1.21	79.0	2.10	250.0	57.0	—	33.0
415	sugarcane, molasses, more than 46% invert sugar more than 79.5 degrees brix (Black strap)	4-04-696	75.0	0.75	2.31	0.32	0.08	2.86	0.16	0.35	0.90	59.0	1.57	186.0	42.0	—	22.0
416			100.0	1.00	3.10	0.43	0.11	3.84	0.22	0.47	1.21	79.0	2.10	250.0	56.0	—	30.0
417	wood, molasses	4-05-502	63.0	1.34	0.12	0.07	0.04	0.04	0.03	0.03	—	—	—	—	13.0	—	—
418			100.0	2.15	0.20	0.11	0.06	0.06	0.05	0.05	—	—	—	—	20.0	—	—
	NAPIERGRASS *Pennisetum purpureum*																	
419	fresh	2-03-166	21.0	0.09	—	0.06	0.07	0.28	0.00	0.02	—	—	—	—	—	—	—
420			100.0	0.44	—	0.26	0.35	1.31	0.01	0.10	—	—	—	—	—	—	—
	NEEDLEAND-THREAD *Stipa comata*																	
421	fresh, stem-cured	2-07-989	92.0	0.99	—	—	0.06	—	—	—	—	—	—	—	—	—	—
422			100.0	1.08	—	—	0.06	—	—	—	—	—	—	—	—	—	—
	OATS *Avena sativa*																	
423	breakfast cereal by-product, less than 4% fiber (Feeding oat meal) (Oat middlings)	4-03-303	91.0	0.07	0.05	0.14	0.44	0.50	0.09	0.22	0.05	4.0	—	382.0	44.0	—	139.0
424			100.0	0.08	0.06	0.16	0.49	0.55	0.10	0.24	0.05	5.0	—	421.0	48.0	—	154.0
425	grain	4-03-309	89.0	0.07	0.09	0.13	0.33	0.39	0.07	0.21	0.06	6.0	0.10	76.0	37.0	0.23	37.0
426			100.0	0.07	0.11	0.14	0.38	0.44	0.08	0.23	0.06	7.0	0.11	85.0	42.0	0.26	41.0
427	grain, Pacific Coast	4-07-999	91.0	0.10	0.12	0.17	0.31	0.38	0.06	0.20	—	—	—	73.0	38.0	0.08	—
428	Coast		100.0	0.11	0.13	0.19	0.34	0.42	0.07	0.22	—	—	—	80.0	42.0	0.08	—
429	groats	4-03-331	90.0	0.08	0.08	0.11	0.43	0.35	0.05	0.20	—	6.0	0.11	73.0	28.0	—	0.0
430			100.0	0.08	0.09	0.13	0.48	0.39	0.06	0.22	—	7.0	0.12	82.0	31.0	—	0.0
431	hay, sun-cured	1-03-280	91.0	0.22	0.48	0.24	0.20	1.38	0.17	0.22	0.07	14.0	—	142.0	59.0	0.16	36.0
432			100.0	0.24	0.52	0.26	0.22	1.51	0.18	0.25	0.07	15.0	—	155.0	64.0	0.17	39.0
433	hulls	1-03-281	92.0	0.14	0.08	0.08	0.14	0.57	0.04	0.14	—	4.0	—	102.0	19.0	—	—
434			100.0	0.15	0.08	0.09	0.15	0.62	0.04	0.15	—	4.0	—	111.0	20.0	—	—
435	silage	3-03-298	31.0	0.10	—	0.09	0.07	0.84	0.07	0.09	0.02	2.0	—	65.0	13.0	—	11.0
436			100.0	0.34	—	0.30	0.24	2.74	0.23	0.29	0.06	6.0	—	211.0	43.0	—	35.0
437	straw	1-03-283	92.0	0.22	0.71	0.17	0.06	2.37	0.39	0.21	—	9.0	—	161.0	34.0	—	6.0
438			100.0	0.24	0.78	0.18	0.06	2.57	0.42	0.23	—	10.0	—	175.0	37.0	—	6.0
	ORANGE *Citrus sinensis*																	
439	pulp without fines, dehy (Orange pulp, dried)	4-01-254	88.0	0.62	—	—	0.10	—	—	—	—	—	—	—	—	—	—
440			100.0	0.71	—	—	0.11	—	—	—	—	—	—	—	—	—	—

TABLE 2 Composition of Important Feeds: Mineral Elements—*Continued*

Entry Number	Feed Name Description	International Feed Number	Dry Matter (%)	Calcium (%)	Chlorine (%)	Magnesium (%)	Phosphorus (%)	Potassium (%)	Sodium (%)	Sulfur (%)	Cobalt (mg/kg)	Copper (mg/kg)	Iodine (mg/kg)	Iron (mg/kg)	Manganese (mg/kg)	Selenium (mg/kg)	Zinc (mg/kg)
	ORCHARDGRASS *Dactylis glomerata*																
441	fresh	2-03-451	27.0	0.10	—	0.08	0.11	0.89	0.01	0.07	—	6.0	—	126.0	28.0	—	—
442			100.0	0.37	—	0.29	0.39	3.33	0.04	0.26	—	24.0	—	470.0	104.0	—	—
443	fresh, early vege-	2-03-439	23.0	0.13	0.02	0.07	0.13	0.84	0.01	0.05	—	2.0	—	39.0	22.0	—	—
444	tative		100.0	0.58	0.08	0.31	0.54	3.58	0.04	0.21	—	7.0	—	169.0	96.0	—	—
445	hay, sun-cured	1-03-438	91.0	0.35	0.37	0.15	0.32	3.06	0.05	0.24	0.42	12.0	—	90.0	109.0	—	23.0
446			100.0	0.39	0.41	0.17	0.35	3.36	0.05	0.26	0.46	13.0	—	99.0	120.0	—	26.0
	PANGOLAGRASS *Digitaria decumbens*																
447	fresh	2-03-493	21.0	0.09	—	0.03	0.04	—	—	—	—	—	—	—	—	—	—
448			100.0	0.43	—	0.14	0.18	—	—	—	—	—	—	—	—	—	—
449	hay, sun-cured	1-10-638	91.0	0.53	—	0.18	0.19	1.55	—	—	—	—	—	—	—	—	—
450	15 to 28 days' growth (South)		100.0	0.58	—	0.20	0.21	1.70	—	—	—	—	—	—	—	—	—
451	hay, sun-cured,	1-26-214	91.0	0.42	—	0.14	0.21	1.27	—	—	—	—	—	—	—	—	—
452	29 to 42 days' growth (South)		100.0	0.46	—	0.15	0.23	1.40	—	—	—	—	—	—	—	—	—
453	hay, sun-cured,	1-29-573	91.0	0.35	—	0.13	0.16	1.00	—	—	—	—	—	—	—	—	—
454	43 to 56 days' growth (South)		100.0	0.38	—	0.14	0.18	1.10	—	—	—	—	—	—	—	—	—
	PEA *Pisum* spp																
455	seeds	5-03-600	89.0	0.14	0.06	0.13	0.39	1.01	0.04	—	—	—	—	50.0	—	—	29.0
456			100.0	0.15	0.06	0.14	0.44	1.13	0.05	—	—	—	—	57.0	—	—	33.0
457	vines without	3-03-596	25.0	0.32	—	0.10	0.06	0.34	0.00	0.06	—	—	—	25.0	—	—	—
458	seeds, silage		100.0	1.31	—	0.39	0.24	1.40	0.01	0.25	—	—	—	100.0	—	—	—
	PEANUT *Arachis hypogaea*																
459	hay, sun-cured	1-03-619	91.0	1.12	—	0.45	0.14	1.25	—	0.21	0.07	—	—	—	—	—	—
460			100.0	1.23	—	0.49	0.15	1.38	—	0.23	0.08	—	—	—	—	—	—
461	kernels, meal mech extd (Peanut meal)	5-03-649	93.0	0.19	0.03	0.29	0.57	1.16	0.21	0.27	0.11	15.0	0.07	156.0	26.0	0.29	21.0
462			100.0	0.20	0.03	0.31	0.61	1.25	0.23	0.29	0.12	16.0	0.07	169.0	28.0	0.31	22.0
463	kernels, meal solv extd (Peanut meal)	5-03-650	92.0	0.27	0.03	0.15	0.62	1.13	0.07	0.30	0.11	15.0	0.07	142.0	27.0	—	20.0
464			100.0	0.29	0.03	0.17	0.68	1.23	0.08	0.33	0.12	17.0	0.07	154.0	29.0	—	22.0
	PINEAPPLE *Ananas comosus*																
465	aerial part without fruit, sun-cured (Pineapple hay)	1-13-309	89.0	0.35	—	—	0.21	—	—	—	—	—	—	—	—	—	—
466			100.0	0.39	—	—	0.23	—	—	—	—	—	—	—	—	—	—
467	process residue, dehy (Pineapple bran)	4-03-722	87.0	0.20	—	—	0.11	—	—	—	—	—	—	489.0	—	—	—
468			100.0	0.23	—	—	0.13	—	—	—	—	—	—	561.0	—	—	—
	POTATO *Solanum tuberosum*																
469	process residue, dehy	4-03-775	89.0	0.14	—	—	0.23	—	—	—	—	—	—	—	—	—	—
470			100.0	0.16	—	—	0.25	—	—	—	—	—	—	—	2.0	—	—
471	tubers, dehy	4-07-850	91.0	0.07	0.36	0.11	0.20	1.96	0.01	0.08	—	—	—	—	2.0	—	2.0
472			100.0	0.08	0.40	0.12	0.22	2.15	0.01	0.09	—	—	—	—	2.0	—	2.0

TABLE 2 Composition of Important Feeds: Mineral Elements—*Continued*

Entry Number	Feed Name Description	International Feed Number	Dry Matter (%)	Calcium (%)	Chlorine (%)	Magnesium (%)	Phosphorus (%)	Potassium (%)	Sodium (%)	Sulfur (%)	Cobalt (mg/kg)	Copper (mg/kg)	Iodine (mg/kg)	Iron (mg/kg)	Manganese (mg/kg)	Selenium (mg/kg)	Zinc (mg/kg)
473	tubers, fresh	4-03-787	23.0	0.01	0.07	0.03	0.06	0.51	0.02	0.02	—	7.0	—	18.0	10.0	—	—
474			100.0	0.04	0.28	0.14	0.24	2.17	0.09	0.09	—	28.0	—	78.0	42.0	—	—
475	tubers, silage	4-03-768	25.0	0.01	—	0.04	0.06	0.53	0.02	0.06	—	—	—	22.0	—	—	—
476			100.0	0.04	—	0.14	0.23	2.13	0.09	0.23	—	—	—	90.0	—	—	—
477	vines, silage	3-03-765	15.0	0.31	0.06	0.02	0.03	0.59	—	0.06	—	—	—	—	—	—	—
478			100.0	2.12	0.38	0.14	0.20	3.95	—	0.37	—	—	—	—	—	—	—
	POULTRY																
479	by-product, meal	5-03-798	93.0	3.51	0.54	0.18	1.83	0.39	0.82	0.52	0.22	14.0	3.09	442.0	11.0	0.78	121.0
480	rendered (Viscera with feet with heads)		100.0	3.76	0.58	0.19	1.96	0.42	0.87	0.56	0.24	15.0	3.31	473.0	12.0	0.83	129.0
481	feathers, hydrolyzed	5-03-795	93.0	0.26	0.28	0.20	0.67	0.29	0.70	1.50	0.04	7.0	0.04	76.0	13.0	0.84	69.0
482			100.0	0.28	0.30	0.22	0.72	0.31	0.76	1.61	0.05	7.0	0.05	81.0	14.0	0.90	74.0
483	manure and litter	5-05-587	89.0	2.82	—	0.45	1.59	1.50	0.45	1.13	—	172.0	—	695.0	258.0	0.70	397.0
484			100.0	3.16	—	0.50	1.78	1.68	0.51	1.26	—	192.0	—	778.0	289.0	0.79	444.0
485	manure, dehy	5-14-015	90.0	8.40	0.86	0.57	2.28	2.03	0.67	0.16	0.00	80.0	—	1,805.0	366.0	—	392.0
486			100.0	9.31	0.95	0.64	2.52	2.25	0.74	0.18	0.00	89.0	—	2,000.0	406.0	—	434.0
	PRAIRIE PLANTS, MIDWEST																
487	hay, sun-cured	1-03-191	92.0	0.39	0.06	0.26	0.14	0.99	0.04	—	0.12	6.0	—	118.0	101.0	—	31.0
488			100.0	0.43	0.06	0.29	0.15	1.08	0.04	—	0.13	7.0	—	129.0	110.0	—	34.0
	PRICKLYPEAR *Opuntia* spp																
489	fresh	2-01-061	17.0	1.61	0.04	0.23	0.02	0.37	0.05	0.04	—	—	—	—	—	—	—
490			100.0	9.61	0.21	1.38	0.12	2.21	0.30	0.23	—	—	—	—	—	—	—
	RAPE *Brassica* spp																
491	fresh	2-03-867	17.0	0.22	0.08	0.01	0.07	0.50	0.01	0.10	—	1.0	—	30.0	8.0	—	—
492			100.0	1.33	0.45	0.07	0.39	2.98	0.05	0.58	—	8.0	—	182.0	46.0	—	—
493	seeds, meal mech	5-03-870	92.0	0.66	—	0.50	1.04	0.83	0.46	—	—	7.0	—	175.0	55.0	0.96	43.0
494	extd		100.0	0.72	—	0.54	1.14	0.90	0.50	—	—	7.0	—	190.0	60.0	1.04	47.0
495	seeds, meal solv	5-03-871	91.0	0.61	0.10	0.55	0.95	1.24	0.09	1.14	—	—	—	—	—	0.97	—
496	extd		100.0	0.67	0.11	0.60	1.04	1.36	0.10	1.25	—	—	—	—	—	1.07	—
	RAPE, SUMMER *Brassica napus*																
497	seeds, meal mech	5-08-136	94.0	0.71	—	—	1.00	—	—	—	—	—	—	—	—	—	—
498	extd		100.0	0.76	—	—	1.06	—	—	—	—	—	—	—	—	—	—
499	seeds, meal pre-	5-08-135	92.0	0.66	—	—	0.93	—	—	—	—	—	—	—	—	—	—
500	pressed solv extd		100.0	0.72	—	—	1.01	—	—	—	—	—	—	—	—	—	—
	REDTOP *Agrostis alba*																
501	fresh	2-03-897	29.0	0.14	0.03	0.07	0.09	0.69	0.02	0.06	—	8.0	—	59.0	—	—	—
502			100.0	0.46	0.09	0.23	0.29	2.35	0.05	0.19	—	26.0	—	200.0	—	—	—
503	hay, sun-cured,	1-03-886	94.0	0.60	—	—	0.33	1.60	—	—	—	—	—	—	—	—	—
504	midbloom		100.0	0.63	—	—	0.35	1.69	—	—	—	—	—	—	—	—	—
	RICE *Oryza sativa*																
505	bran with germ	4-03-928	91.0	0.07	0.07	0.94	1.54	1.74	0.03	0.18	—	13.0	—	190.0	376.0	0.40	29.0
506	(Rice bran)		100.0	0.08	0.08	1.04	1.70	1.92	0.04	0.20	—	15.0	—	210.0	415.0	0.44	32.0

TABLE 2   Composition of Important Feeds: Mineral Elements—*Continued*

Entry Number	Feed Name Description	International Feed Number	Dry Matter (%)	Calcium (%)	Chlorine (%)	Magnesium (%)	Phosphorus (%)	Potassium (%)	Sodium (%)	Sulfur (%)	Cobalt (mg/kg)	Copper (mg/kg)	Iodine (mg/kg)	Iron (mg/kg)	Manganese (mg/kg)	Selenium (mg/kg)	Zinc (mg/kg)
507	grain, ground	4-03-938	89.0	0.06	0.08	0.13	0.28	0.32	0.05	0.05	0.04	3.0	0.04	51.0	18.0	—	15.0
508	(Ground rough rice) (Ground paddy rice)		100.0	0.07	0.09	0.15	0.32	0.36	0.06	0.05	0.05	3.0	0.05	57.0	20.0	—	17.0
509	grain, polished	4-03-932	89.0	0.03	0.08	0.11	0.27	0.13	0.07	0.04	—	—	—	—	18.0	0.27	17.0
510	and broken (Brewers rice)		100.0	0.03	0.09	0.12	0.30	0.15	0.08	0.05	—	—	—	—	20.0	0.31	19.0
511	groats, polished	4-03-942	89.0	0.02	0.04	0.02	0.11	0.11	0.02	0.08	—	3.0	—	14.0	11.0	—	2.0
512	(Rice, polished)		100.0	0.03	0.04	0.02	0.13	0.12	0.02	0.09	—	3.0	—	16.0	12.0	—	2.0
513	hulls	1-08-075	92.0	0.09	0.07	0.76	0.07	0.52	0.11	0.08	—	—	—	—	308.0	—	—
514			100.0	0.10	0.08	0.83	0.08	0.57	0.12	0.09	—	—	—	—	334.0	—	—
515	polishings	4-03-943	90.0	0.05	0.11	0.78	1.33	1.14	0.10	0.17	—	3.0	—	161.0	12.0	—	26.0
516			100.0	0.05	0.12	0.87	1.48	1.27	0.12	0.19	—	4.0	—	178.0	14.0	—	29.0
	RUSSIANTHISTLE, TUMBLING *Salsola kali tenuifolia*																	
517	fresh, stem cured	2-08-000	88.0	2.91	—	—	0.11	—	—	—	—	—	—	—	—	—	—	—
518			100.0	3.31	—	—	0.12	—	—	—	—	—	—	—	—	—	—
519	hay, sun-cured	1-03-988	86.0	1.41	—	0.77	0.19	5.88	—	0.36	—	—	—	—	—	—	—
520			100.0	1.64	—	0.89	0.22	6.85	—	0.42	—	—	—	—	—	—	—
	RYE *Secale cereale*																	
521	distillers grains, dehy	5-04-023	92.0	0.15	0.05	0.17	0.48	0.07	0.17	0.44	—	—	—	—	18.0	—	—
522			100.0	0.16	0.05	0.18	0.52	0.08	0.18	0.48	—	—	—	—	20.0	—	—
523	fresh	2-04-018	24.0	0.09	—	0.08	0.08	0.82	0.02	—	—	—	—	—	—	—	—
524			100.0	0.39	—	0.31	0.33	3.40	0.07	—	—	—	—	—	—	—	—
525	grain	4-04-047	88.0	0.06	0.03	0.12	0.32	0.46	0.02	0.15	—	7.0	—	60.0	58.0	0.38	31.0
526			100.0	0.07	0.03	0.14	0.37	0.52	0.03	0.17	—	8.0	—	69.0	66.0	0.44	36.0
527	flour by-product, less than 4.5% fiber (Rye middlings)	4-04-032	89.0	0.36	—	0.27	0.10	1.10	0.02	—	—	—	—	—	178.0	—	—
528			100.0	0.40	—	0.30	0.11	1.24	0.02	—	—	—	—	—	200.0	—	—
529	flour by-product, less than 8.5% fiber (Rye middlings)	4-04-031	89.0	0.06	—	—	0.62	0.62	—	—	—	—	—	—	44.0	—	—
530			100.0	0.07	—	—	0.70	0.70	—	—	—	—	—	—	49.0	—	—
531	mill run, less than 9.5% fiber (Rye feed)	4-04-034	90.0	0.08	—	0.23	0.64	0.83	—	0.04	—	—	—	—	—	—	—
532			100.0	0.08	—	0.26	0.71	0.92	—	0.04	—	—	—	—	—	—	—
533	silage	3-04-020	32.0	0.13	—	—	0.10	—	—	—	—	—	—	—	—	—	—
534			100.0	0.39	—	—	0.32	—	—	—	—	—	—	—	—	—	—
535	straw	1-04-007	90.0	0.22	0.21	0.07	0.08	0.87	0.12	0.10	—	4.0	—	—	6.0	—	—
536			100.0	0.24	0.24	0.08	0.09	0.97	0.13	0.11	—	4.0	—	—	7.0	—	—
	RYEGRASS, ITALIAN *Lolium multiflorum*																	
537	fresh	2-04-073	25.0	0.16	—	0.09	0.10	0.49	0.00	0.03	—	—	—	160.0	—	—	—
538			100.0	0.65	—	0.35	0.41	2.00	0.01	0.10	—	—	—	650.0	—	—	—
539	hay, sun-cured	1-04-069	86.0	0.53	—	0.28	0.29	1.34	—	—	—	—	—	275.0	—	—	—
540			100.0	0.62	—	0.32	0.34	1.56	—	—	—	—	—	320.0	—	—	—

TABLE 2 Composition of Important Feeds: Mineral Elements—*Continued*

Entry Num-ber	Feed Name Description	International Feed Number	Dry Mat-ter (%)	Cal-cium (%)	Chlo-rine (%)	Mag-ne-sium (%)	Phos-pho-rus (%)	Potas-sium (%)	So-dium (%)	Sul-fur (%)	Co-balt (mg/kg)	Cop-per (mg/kg)	Io-dine (mg/kg)	Iron (mg/kg)	Man-ga-nese (mg/kg)	Sele-nium (mg/kg)	Zinc (mg/kg)
	RYEGRASS, PE-RENNIAL *Lolium perenne*																
541	fresh	2-04-086	27.0	0.15	—	—	0.07	0.51	—	0.08	0.02	3.0	—	—	—	—	—
542			100.0	0.55	—	—	0.27	1.91	—	0.30	0.06	13.0	—	—	—	—	—
543	hay, sun-cured	1-04-077	86.0	0.56	—	—	0.28	1.44	—	—	—	—	—	—	—	—	—
544			100.0	0.65	—	—	0.32	1.67	—	—	—	—	—	—	—	—	—
	SAFFLOWER *Carthamus tinctorius*																
545	seeds	4-07-958	94.0	0.24	—	0.34	0.63	0.74	0.06	—	—	10.0	—	468.0	19.0	—	—
546			100.0	0.26	—	0.36	0.67	0.79	0.06	—	—	11.0	—	500.0	20.0	—	44.0
547	seeds, meal mech	5-04-109	91.0	0.25	—	0.33	0.71	0.72	0.05	—	—	10.0	—	471.0	18.0	—	40.0
548	extd		100.0	0.27	—	0.36	0.78	0.79	0.05	—	—	11.0	—	515.0	20.0	—	44.0
549	seeds, meal solv	5-04-110	92.0	0.34	—	0.35	0.75	0.76	0.05	0.13	—	10.0	—	495.0	18.0	—	41.0
550	extd		100.0	0.37	—	0.37	0.81	0.82	0.05	0.14	—	11.0	—	537.0	20.0	—	44.0
551	seeds without hulls, meal solv extd	5-07-959	92.0	0.35	0.16	1.02	1.29	1.10	0.04	0.20	1.97	9.0	—	484.0	39.0	—	33.0
552			100.0	0.38	0.18	1.11	1.40	1.19	0.05	0.22	2.15	9.0	—	528.0	43.0	—	36.0
	SAGE, BLACK *Salvia mellifera*																
553	browse, fresh,	2-05-564	65.0	0.53	—	—	0.11	—	—	—	—	—	—	—	—	—	—
554	stem-cured		100.0	0.81	—	—	0.17	—	—	—	—	—	—	—	—	—	—
	SAGEBRUSH, BIG *Artemisia tridentata*																
555	browse, fresh,	2-07-992	65.0	0.46	—	—	0.12	—	—	—	—	—	—	—	—	—	—
556	stem-cured		100.0	0.71	—	—	0.18	—	—	—	—	—	—	—	—	—	—
	SAGEBRUSH, BUD *Artemisia spinescens*																
557	browse, fresh,	2-07-991	23.0	0.22	—	—	0.08	—	—	—	—	—	—	—	—	—	—
558	early vegetative		100.0	0.97	—	—	0.33	—	—	—	—	—	—	—	—	—	—
559	browse, fresh,	2-04-124	32.0	0.19	—	0.16	0.13	—	—	—	—	—	—	—	—	—	—
560	late vegetative		100.0	0.60	—	0.49	0.42	—	—	—	—	—	—	—	—	—	—
	SALTBUSH, NUT-TALL *Atriplex nuttallii*																
561	browse, fresh,	2-07-993	55.0	1.22	—	—	0.06	—	—	—	—	—	—	—	—	—	—
562	stem-cured		100.0	2.21	—	—	0.12	—	—	—	—	—	—	—	—	—	—
	SALTGRASS *Distichlis* spp																
563	fresh	2-04-170	74.0	0.16	—	0.22	0.06	0.18	—	—	—	—	—	141.0	115.0	—	—
564			100.0	0.22	—	0.30	0.08	0.24	—	—	—	—	—	190.0	155.0	—	—
	SALTGRASS, DES-ERT *Distichlis stricta*																
565	fresh	2-04-171	29.0	0.05	—	—	0.03	—	—	—	—	—	—	—	—	—	—
566			100.0	0.16	—	—	0.09	—	—	—	—	—	—	—	—	—	—

TABLE 2   Composition of Important Feeds: Mineral Elements—*Continued*

Entry Number	Feed Name Description	International Feed Number	Dry Matter (%)	Calcium (%)	Chlorine (%)	Magnesium (%)	Phosphorus (%)	Potassium (%)	Sodium (%)	Sulfur (%)	Cobalt (mg/kg)	Copper (mg/kg)	Iodine (mg/kg)	Iron (mg/kg)	Manganese (mg/kg)	Selenium (mg/kg)	Zinc (mg/kg)
	SCREENINGS—SEE BARLEY, SEE CEREALS, SEE WHEAT																
	SEAWEED, KELP *Laminariales* (order)–*Fucales* (order)																
567	whole, dehy	4-08-073	91.0	2.48	—	0.85	0.28	—	—	—	—	—	—	—	—	—	—
568			100.0	2.72	—	0.93	0.31	—	—	—	—	—	—	—	—	—	—
	SESAME *Sesamum indicum*																
569	seeds, meal mech	5-04-220	93.0	2.01	0.07	0.46	1.36	1.25	0.04	0.33	—	—	—	93.0	48.0	—	100.0
570	extd		100.0	2.17	0.07	0.50	1.46	1.35	0.04	0.35	—	—	—	100.0	52.0	—	108.0
	SHRIMP *Pandalus* spp–*Penaeus* spp																
571	process residue,	5-04-226	90.0	9.73	1.04	0.54	1.84	0.83	1.57	—	—	—	—	105.0	30.0	—	28.0
572	meal (Shrimp meal)		100.0	10.80	1.15	0.60	2.05	0.92	1.74	—	—	—	—	116.0	33.0	—	32.0
	SORGHUM *Sorghum bicolor*																
573	aerial part with	1-07-960	89.0	0.35	—	0.26	0.18	1.31	0.02	—	—	—	—	—	—	—	—
574	heads, sun-cured (Fodder)		100.0	0.40	—	0.29	0.21	1.47	0.02	—	—	—	—	—	—	—	—
575	aerial part without	1-04-302	88.0	0.46	—	0.25	0.12	1.06	0.02	—	—	—	—	—	—	—	—
576	heads, sun-cured (Stover)		100.0	0.52	—	0.28	0.13	1.20	0.02	—	—	—	—	—	—	—	—
577	distillers grains,	5-04-374	94.0	0.15	—	0.18	0.69	0.36	0.05	0.17	—	—	—	47.0	—	—	—
578	dehy		100.0	0.16	—	0.19	0.74	0.38	0.05	0.18	—	—	—	50.0	—	—	—
579	grain	4-04-383	90.0	0.03	0.09	0.16	0.29	0.35	0.03	0.13	0.16	10.0	0.04	45.0	16.0	0.44	17.0
580			100.0	0.04	0.10	0.18	0.33	0.39	0.03	0.15	0.18	11.0	0.04	51.0	18.0	0.50	19.0
581	hay, sun-cured,	1-04-299	92.0	0.46	—	0.46	0.17	2.39	—	—	—	—	—	—	—	—	—
582	early vegetative (South)		100.0	0.50	—	0.50	0.18	2.60	—	—	—	—	—	—	—	—	—
583	hay, sun-cured,	1-06-141	92.0	0.37	—	0.32	0.18	1.75	—	—	—	—	—	—	—	—	—
584	late vegetative (South)		100.0	0.40	—	0.35	0.19	1.90	—	—	—	—	—	—	—	—	—
585	hay, sun-cured,	1-06-142	93.0	0.28	—	0.19	0.13	1.12	—	—	—	—	—	—	—	—	—
586	early bloom (South)		100.0	0.30	—	0.20	0.14	1.20	—	—	—	—	—	—	—	—	—
587	silage	3-04-323	30.0	0.10	0.04	0.09	0.06	0.40	0.01	0.03	0.09	10.0	—	84.0	22.0	—	9.0
588			100.0	0.35	0.13	0.29	0.21	1.37	0.02	0.11	0.30	35.0	—	285.0	73.0	—	32.0
	SORGHUM, JOHNSONGRASS *Sorghum halepense*																
589	hay, sun-cured	1-04-407	89.0	0.75	—	0.31	0.25	1.21	0.01	0.09	—	—	—	527.0	—	—	—
590			100.0	0.84	—	0.35	0.28	1.35	0.01	0.10	—	—	—	590.0	—	—	—
	SORGHUM, SORGO *Sorghum bicolor saccharatum*																
591	silage	3-04-468	27.0	0.09	0.02	0.08	0.05	0.31	0.04	0.03	—	9.0	—	54.0	17.0	—	—
592			100.0	0.34	0.06	0.27	0.17	1.12	0.15	0.10	—	31.0	—	198.0	61.0	—	—

TABLE 2 Composition of Important Feeds: Mineral Elements—*Continued*

Entry Number	Feed Name Description	International Feed Number	Dry Matter (%)	Calcium (%)	Chlorine (%)	Magnesium (%)	Phosphorus (%)	Potassium (%)	Sodium (%)	Sulfur (%)	Cobalt (mg/kg)	Copper (mg/kg)	Iodine (mg/kg)	Iron (mg/kg)	Manganese (mg/kg)	Selenium (mg/kg)	Zinc (mg/kg)
	SORGHUM, SUDAN-GRASS *Sorghum bicolor sudanense*																
593	fresh, early vege-	2-04-484	18.0	0.08	—	0.06	0.07	0.38	0.00	0.02	—	—	—	36.0	—	—	—
594	tative		100.0	0.43	—	0.35	0.41	2.14	0.01	0.11	—	—	—	200.0	—	—	—
595	fresh, midbloom	2-04-485	23.0	0.10	—	0.08	0.08	0.49	0.00	0.03	—	—	—	46.0	—	—	—
596			100.0	0.43	—	0.35	0.36	2.14	0.01	0.11	—	—	—	200.0	—	—	—
597	hay, sun-cured	1-04-480	91.0	0.50	—	0.47	0.28	1.70	0.02	0.06	0.12	34.0	—	176.0	83.0	—	—
598			100.0	0.55	—	0.51	0.30	1.87	0.02	0.06	0.13	37.0	—	193.0	91.0	—	—
599	silage	3-04-499	28.0	0.13	—	0.13	0.06	0.64	0.01	0.02	0.09	11.0	—	36.0	28.0	—	—
600			100.0	0.46	—	0.44	0.21	2.25	0.02	0.06	0.31	37.0	—	127.0	99.0	—	—
	SOYBEAN *Glycine max*																
601	flour by-product	4-04-594	90.0	0.44	—	0.32	0.16	1.51	0.25	0.06	—	—	—	—	29.0	—	—
602	(Soybean mill feed)		100.0	0.49	—	0.36	0.18	1.69	0.28	0.07	—	—	—	—	32.0	—	—
603	fresh, dough stage	2-04-573	26.0	0.34	—	0.21	0.08	0.20	—	—	—	—	—	—	—	—	—
604			100.0	1.31	—	0.83	0.31	0.79	—	—	—	—	—	—	—	—	—
605	hay, sun-cured	1-04-558	89.0	1.15	0.13	0.71	0.25	0.96	0.08	0.21	0.08	8.0	0.22	261.0	95.0	—	22.0
606			100.0	1.29	0.15	0.79	0.28	1.07	0.09	0.24	0.09	9.0	0.24	292.0	106.0	—	24.0
607	hulls (Soybean	1-04-560	91.0	0.45	—	—	0.19	1.16	0.01	0.08	0.11	16.0	—	295.0	10.0	—	22.0
608	flakes) oil—see Fats and oils		100.0	0.49	—	—	0.21	1.27	0.01	0.09	0.12	18.0	—	324.0	11.0	—	24.0
609	protein concen-	5-08-038	92.0	0.11	0.02	0.02	0.68	0.17	0.07	0.70	0.39	14.0	0.32	137.0	5.0	0.14	34.0
610	trate, more than 70% protein		100.0	0.12	0.02	0.02	0.74	0.19	0.08	0.76	0.42	15.0	0.35	149.0	6.0	0.15	37.0
611	seeds	5-04-610	92.0	0.25	0.03	0.26	0.60	1.66	0.02	0.22	—	18.0	—	84.0	36.0	0.11	57.0
612			100.0	0.27	0.03	0.29	0.65	1.82	0.02	0.24	—	20.0	—	91.0	39.0	0.12	62.0
613	seeds, heat pro-	5-04-597	90.0	0.25	—	0.21	0.59	1.70	0.03	0.22	—	16.0	—	80.0	30.0	0.11	54.0
614	cessed		100.0	0.28	—	0.23	0.66	1.89	0.03	0.24	—	18.0	—	89.0	33.0	0.12	60.0
615	seeds, meal mech	5-04-600	90.0	0.26	0.07	0.25	0.61	1.79	0.03	0.33	0.18	22.0	—	157.0	31.0	0.10	60.0
616	extd		100.0	0.29	0.08	0.28	0.68	1.98	0.03	0.37	0.20	24.0	—	175.0	35.0	0.11	66.0
617	seeds, meal solv	5-04-604	90.0	0.30	0.04	0.27	0.63	1.97	0.04	0.43	0.09	23.0	0.13	119.0	29.0	0.30	43.0
618	extd		100.0	0.34	0.04	0.30	0.70	2.20	0.04	0.47	0.10	25.0	0.15	133.0	32.0	0.34	48.0
619	seeds without	5-04-612	90.0	0.26	0.04	0.28	0.63	2.07	0.03	0.44	0.07	20.0	0.11	133.0	37.0	0.10	55.0
620	hulls, meal solv extd		100.0	0.29	0.05	0.32	0.70	2.30	0.03	0.48	0.07	22.0	0.12	148.0	41.0	0.11	61.0
621	silage	3-04-581	27.0	0.37	—	0.10	0.13	0.25	0.03	0.08	—	3.0	—	110.0	31.0	—	—
622			100.0	1.36	—	0.38	0.47	0.93	0.09	0.30	—	9.0	—	400.0	114.0	—	—
623	straw	1-04-567	88.0	1.40	—	0.81	0.05	0.49	0.11	0.23	—	—	—	263.0	45.0	—	—
624			100.0	1.59	—	0.92	0.06	0.56	0.12	0.26	—	—	—	300.0	51.0	—	—
	SPELT *Triticum spelta*																
625	grain	4-04-651	90.0	0.12	—	—	0.38	—	—	—	—	—	—	—	—	—	—
626			100.0	0.13	—	—	0.42	—	—	—	—	—	—	—	—	—	—
	SQUIRRELTAIL *Sitanion* spp																
627	fresh, stem-cured	2-05-566	50.0	0.19	—	—	0.03	—	—	—	—	—	—	—	—	—	—
628			100.0	0.37	—	—	0.06	—	—	—	—	—	—	—	—	—	—
	SUDANGRASS—SEE SORGHUM, SUDANGRASS																

TABLE 2　Composition of Important Feeds: Mineral Elements—*Continued*

Entry Number	Feed Name Description	International Feed Number	Dry Matter (%)	Calcium (%)	Chlorine (%)	Magnesium (%)	Phosphorus (%)	Potassium (%)	Sodium (%)	Sulfur (%)	Cobalt (mg/kg)	Copper (mg/kg)	Iodine (mg/kg)	Iron (mg/kg)	Manganese (mg/kg)	Selenium (mg/kg)	Zinc (mg/kg)
	SUGARCANE *Saccharum officinarum*																
629	bagasse, dehy	1-04-686	91.0	0.82	—	0.09	0.27	0.46	0.18	0.09	—	—	—	91.0	—	—	—
630			100.0	0.90	—	0.10	0.29	0.50	0.20	0.10	—	—	—	100.0	—	—	—
	molasses—see Molasses and syrup																
	SUNFLOWER, COMMON *Helianthus annuus*																
631	seeds, meal solv	5-09-340	90.0	0.21	—	0.68	0.93	0.96	—	0.30	—	—	—	—	—	—	—
632	extd		100.0	0.23	—	0.75	1.03	1.06	—	0.33	—	—	—	—	—	—	—
633	seeds without	5-04-738	93.0	0.39	0.19	0.72	1.06	1.06	0.22	—	—	4.0	—	31.0	21.0	—	—
634	hulls, meal mech extd		100.0	0.42	0.20	0.78	1.14	1.14	0.24	—	—	4.0	—	33.0	22.0	—	—
635	seeds without	5-04-739	93.0	0.41	0.10	0.71	0.91	1.06	0.22	—	—	4.0	—	31.0	19.0	—	—
636	hulls, meal solv extd		100.0	0.44	0.11	0.77	0.98	1.14	0.24	—	—	4.0	—	33.0	20.0	—	—
	SWEETCLOVER, YELLOW *Melilotus officinalis*																
637	hay, sun-cured	1-04-754	87.0	1.11	0.32	0.43	0.22	1.40	0.08	0.41	—	9.0	—	133.0	94.0	—	—
638			100.0	1.27	0.37	0.49	0.25	1.60	0.09	0.47	—	10.0	—	152.0	108.0	—	—
	SWINE *Sus scrofa*																
639	livers, fresh	5-04-792	30.0	0.01	—	0.01	0.37	0.26	0.07	—	0.25	56.0	0.34	145.0	2.0	0.34	44.0
640			100.0	0.04	—	0.04	1.22	0.85	0.24	—	0.84	187.0	1.12	480.0	6.0	1.12	146.0
641	lungs, fresh	5-26-140	16.0	0.01	—	0.01	0.17	0.06	0.15	—	0.33	0.0	0.13	75.0	—	—	11.0
642			100.0	0.05	—	0.04	1.05	0.39	0.96	—	2.09	0.0	0.82	475.0	—	—	68.0
	TIMOTHY *Phleum pratense*																
643	fresh	2-04-912	30.0	0.10	0.17	0.04	0.08	0.59	0.04	0.04	0.01	3.0	—	55.0	43.0	—	—
644			100.0	0.33	0.57	0.14	0.28	1.94	0.14	0.13	0.04	11.0	—	184.0	144.0	—	—
645	hay, sun-cured	1-04-893	90.0	0.39	0.46	0.12	0.18	1.37	0.06	0.16	0.07	5.0	0.03	163.0	45.0	—	15.0
646			100.0	0.43	0.51	0.13	0.20	1.52	0.07	0.17	0.08	5.0	0.04	180.0	50.0	—	17.0
647	hay, sun-cured,	1-04-881	89.0	0.59	—	0.13	0.30	1.50	0.16	—	—	—	—	179.0	—	—	—
648	late vegetative		100.0	0.66	—	0.14	0.34	1.68	0.18	—	—	—	—	200.0	—	—	—
649	hay, sun-cured,	1-04-882	90.0	0.48	—	0.13	0.22	—	0.16	—	—	—	—	179.0	—	—	—
650	early bloom		100.0	0.53	—	0.14	0.25	—	0.18	—	—	—	—	200.0	—	—	—
651	hay, sun-cured,	1-04-883	89.0	0.43	—	0.14	0.20	1.41	0.16	—	—	5.0	—	151.0	—	—	—
652	midbloom		100.0	0.48	—	0.16	0.22	1.59	0.18	—	—	5.0	—	170.0	—	—	—
653	hay, sun-cured,	1-04-884	89.0	0.38	0.55	0.12	0.18	1.45	0.16	—	—	4.0	—	139.0	—	—	—
654	full bloom		100.0	0.43	0.62	0.14	0.20	1.64	0.18	—	—	5.0	—	157.0	—	—	—
655	hay, sun-cured,	1-04-885	88.0	0.34	—	0.12	0.16	1.42	0.06	—	—	—	—	141.0	—	—	—
656	late bloom		100.0	0.38	—	0.13	0.18	1.61	0.07	—	—	—	—	160.0	—	—	—
657	hay, sun-cured,	1-04-886	92.0	0.26	—	0.11	0.17	—	0.01	—	—	—	—	—	—	—	—
658	milk stage		100.0	0.28	—	0.12	0.18	—	0.01	—	—	—	—	—	—	—	—
659	silage	3-04-922	34.0	0.19	—	0.05	0.10	0.58	0.04	0.04	—	2.0	—	37.0	31.0	—	—
660			100.0	0.55	—	0.15	0.29	1.69	0.11	0.13	—	6.0	—	110.0	90.0	—	—
	TOMATO *Lycopersicon esculentum*																
661	pomace, dehy	5-05-041	92.0	0.39	—	0.18	0.55	3.33	—	—	—	30.0	—	4,223.0	47.0	—	—
662			100.0	0.43	—	0.20	0.60	3.63	—	—	—	33.0	—	4,600.0	51.0	—	—

TABLE 2 Composition of Important Feeds: Mineral Elements—*Continued*

Entry Number	Feed Name Description	International Feed Number	Dry Matter (%)	Calcium (%)	Chlorine (%)	Magnesium (%)	Phosphorus (%)	Potassium (%)	Sodium (%)	Sulfur (%)	Cobalt (mg/kg)	Copper (mg/kg)	Iodine (mg/kg)	Iron (mg/kg)	Manganese (mg/kg)	Selenium (mg/kg)	Zinc (mg/kg)
	TORULA DRIED YEAST— SEE YEAST, TORULA																
	TREFOIL, BIRDS-FOOT *Lotus corniculatus*																
663	fresh	2-20-786	24.0	0.46	—	0.07	0.05	0.48	0.02	0.06	0.05	—	—	97.0	—	—	—
664			100.0	1.91	—	0.28	0.22	1.99	0.07	0.25	0.21	—	—	400.0	—	—	—
665	hay, sun-cured	1-05-044	92.0	1.57	—	0.47	0.25	1.77	0.06	0.23	0.10	9.0	—	210.0	26.0	—	—
666			100.0	1.70	—	0.51	0.27	1.92	0.07	0.25	0.11	9.0	—	228.0	29.0	—	—
	TRITICALE *Triticale hexaploide*																
667	grain	4-20-362	90.0	0.05	—	—	0.30	0.36	—	0.15	—	—	—	—	—	—	—
668			100.0	0.06	—	—	0.33	0.40	—	0.17	—	—	—	—	—	—	—
	TURNIP *Brassica rapa rapa*																
669	roots, fresh	4-05-067	9.0	0.05	0.06	0.02	0.02	0.28	0.10	0.04	—	2.0	—	11.0	4.0	—	—
670			100.0	0.59	0.65	0.22	0.26	2.99	1.05	0.43	—	21.0	—	118.0	43.0	—	—
	VETCH *Vicia* spp																
671	hay, sun-cured	1-05-106	89.0	1.05	—	0.22	0.29	2.07	0.46	0.13	0.32	9.0	0.44	374.0	65.0	—	—
672			100.0	1.18	—	0.25	0.32	2.32	0.52	0.15	0.36	10.0	0.49	420.0	73.0	—	—
	WHALE *Balaena glacialis–Balaenoptera* spp																
673	meat, meal rendered	5-05-160	91.0	0.40	—	—	0.56	—	—	—	—	—	—	—	—	—	—
674			100.0	0.44	—	—	0.61	—	—	—	—	—	—	—	—	—	—
	WHEAT *Triticum aestivum*																
675	bran	4-05-190	89.0	0.11	0.05	0.53	1.22	1.38	0.04	0.22	0.10	13.0	0.07	114.0	111.0	0.38	114.0
676			100.0	0.13	0.05	0.60	1.38	1.56	0.04	0.25	0.11	14.0	0.07	128.0	125.0	0.43	128.0
677	bread, dehy	4-07-944	95.0	0.06	—	—	0.11	—	—	—	—	—	—	—	—	—	—
678			100.0	0.07	—	—	0.11	—	—	—	—	—	—	—	—	—	—
679	flour, hard red spring, less than 1.5% fiber	4-08-113	88.0	0.04	0.10	0.09	0.25	0.21	0.01	0.22	0.06	1.0	0.09	4.0	4.0	0.30	6.0
680			100.0	0.04	0.11	0.11	0.28	0.23	0.01	0.25	0.07	1.0	0.10	4.0	4.0	0.34	7.0
681	flour, less than 1.5% fiber (Wheat feed flour)	4-05-199	88.0	0.03	0.09	0.05	0.18	0.14	0.01	0.21	0.06	1.0	0.09	5.0	3.0	0.15	6.0
682			100.0	0.03	0.10	0.06	0.20	0.16	0.01	0.24	0.07	1.0	0.10	6.0	4.0	0.17	7.0
683	flour by-product, less than 4% fiber (Wheat red dog)	4-05-203	88.0	0.04	0.14	0.16	0.49	0.51	0.04	0.24	0.12	6.0	—	46.0	55.0	0.30	65.0
684			100.0	0.05	0.16	0.18	0.56	0.58	0.05	0.27	0.13	7.0	—	52.0	62.0	0.35	74.0
685	flour by-product, less than 7% fiber (Wheat shorts)	4-05-201	88.0	0.09	0.07	0.25	0.81	0.93	0.02	0.20	0.10	12.0	—	73.0	117.0	0.43	109.0
686			100.0	0.10	0.08	0.28	0.91	1.06	0.03	0.22	0.12	13.0	—	82.0	132.0	0.49	124.0
687	flour by-product, less than 9.5% fiber (Wheat middlings)	4-05-205	89.0	0.11	0.04	0.36	0.88	1.00	0.17	0.17	0.09	19.0	0.11	83.0	112.0	0.74	103.0
688			100.0	0.13	0.04	0.40	0.99	1.13	0.19	0.20	0.10	22.0	0.12	93.0	126.0	0.83	116.0

TABLE 2　Composition of Important Feeds: Mineral Elements—*Continued*

Entry Number	Feed Name Description	International Feed Number	Dry Matter (%)	Calcium (%)	Chlorine (%)	Magnesium (%)	Phosphorus (%)	Potassium (%)	Sodium (%)	Sulfur (%)	Cobalt (mg/kg)	Copper (mg/kg)	Iodine (mg/kg)	Iron (mg/kg)	Manganese (mg/kg)	Selenium (mg/kg)	Zinc (mg/kg)
689	mill run, less	4-05-206	90.0	0.10	—	0.47	1.02	1.19	0.22	0.30	0.21	18.0	—	95.0	104.0	—	—
690	than 9.5% fiber		100.0	0.11	—	0.52	1.13	1.33	0.24	0.34	0.23	21.0	—	105.0	116.0	—	—
691	fresh	2-08-078	25.0	0.07	0.16	0.05	0.08	0.67	0.03	0.05	0.02	—	—	25.0	—	—	—
692			100.0	0.28	0.67	0.20	0.31	2.74	0.11	0.22	0.08	—	—	100.0	—	—	—
693	germs, ground	5-05-218	88.0	0.05	0.07	0.25	0.92	0.97	0.03	0.25	0.12	10.0	—	51.0	134.0	0.36	119.0
694			100.0	0.06	0.08	0.28	1.05	1.09	0.03	0.28	0.13	11.0	—	58.0	151.0	0.40	135.0
695	grain	4-05-211	89.0	0.04	0.07	0.15	0.37	0.38	0.04	0.16	0.12	6.0	0.09	54.0	37.0	0.26	44.0
696			100.0	0.04	0.08	0.16	0.42	0.42	0.05	0.18	0.14	7.0	0.10	61.0	42.0	0.30	50.0
697	grain, hard red	4-05-258	88.0	0.03	0.08	0.15	0.38	0.36	0.02	0.15	0.12	6.0	—	56.0	37.0	0.25	45.0
698	spring		100.0	0.04	0.09	0.17	0.43	0.41	0.03	0.17	0.13	7.0	—	64.0	42.0	0.29	52.0
699	grain, hard red	4-05-268	88.0	0.04	0.05	0.11	0.38	0.43	0.02	0.13	0.14	5.0	—	31.0	29.0	0.40	38.0
700	winter		100.0	0.05	0.06	0.13	0.43	0.49	0.02	0.15	0.16	5.0	—	35.0	33.0	0.45	43.0
701	grain, soft red	4-05-294	88.0	0.04	0.07	0.10	0.38	0.41	0.01	0.11	0.10	6.0	—	27.0	32.0	0.04	42.0
702	winter		100.0	0.05	0.08	0.11	0.43	0.46	0.01	0.12	0.12	7.0	—	30.0	36.0	0.05	48.0
703	grain, soft white	4-05-337	89.0	0.06	0.08	0.11	0.32	0.41	0.04	0.14	0.13	7.0	—	36.0	38.0	0.05	26.0
704	winter		100.0	0.07	0.09	0.13	0.36	0.46	0.04	0.16	0.15	8.0	—	41.0	43.0	0.06	29.0
705	grain, soft white	4-08-555	89.0	0.09	—	0.13	0.30	0.45	0.09	0.16	—	—	—	54.0	—	—	—
706	winter, Pacific Coast		100.0	0.10	—	0.15	0.34	0.51	0.10	0.18	—	—	—	60.0	—	—	—
707	grain screenings	4-05-216	89.0	0.13	—	0.16	0.35	0.52	0.09	0.20	—	—	—	54.0	29.0	0.61	—
708			100.0	0.15	—	0.18	0.39	0.58	0.10	0.22	—	—	—	60.0	33.0	0.68	—
709	grits	4-07-852	90.0	0.03	—	—	0.11	0.08	0.00	—	—	—	—	15.0	—	—	—
710			100.0	0.03	—	—	0.12	0.09	0.00	—	—	—	—	17.0	—	—	—
711	hay, sun-cured	1-05-172	88.0	0.13	—	0.11	0.17	0.87	0.18	0.19	—	—	—	175.0	—	—	—
712			100.0	0.15	—	0.12	0.20	1.00	0.21	0.22	—	—	—	200.0	—	—	—
713	silage, early vege-	3-05-184	30.0	0.08	0.02	0.19	0.08	0.42	0.02	0.07	0.01	4.0	—	57.0	39.0	—	—
714	tative		100.0	0.27	0.07	0.62	0.27	1.39	0.07	0.24	0.05	14.0	—	190.0	130.0	—	—
715	straw	1-05-175	89.0	0.16	0.28	0.11	0.04	1.26	0.13	0.17	0.04	3.0	—	140.0	36.0	—	6.0
716			100.0	0.18	0.32	0.12	0.05	1.42	0.14	0.19	0.05	4.0	—	157.0	41.0	—	6.0
	WHEAT, DURUM *Triticum durum*																	
717	grain	4-05-224	88.0	0.09	—	0.15	0.36	0.44	—	—	0.08	7.0	—	42.0	28.0	0.89	33.0
718			100.0	0.10	—	0.17	0.41	0.51	—	—	0.09	8.0	—	48.0	32.0	1.02	37.0
	WHEATGRASS, CRESTED *Agropyron desertorum*																	
719	fresh	2-05-429	39.0	0.18	—	0.11	0.07	—	—	—	—	—	—	—	—	—	—
720			100.0	0.45	—	0.28	0.19	—	—	—	—	—	—	—	—	—	—
721	fresh, early vege-	2-05-420	28.0	0.13	—	0.08	0.10	—	—	—	—	—	—	—	—	—	—
722	tative		100.0	0.46	—	0.28	0.34	—	—	—	—	—	—	—	—	—	—
723	fresh, full bloom	2-05-424	45.0	0.18	—	—	0.13	—	—	—	—	—	—	—	—	—	—
724			100.0	0.39	—	—	0.28	—	—	—	—	—	—	—	—	—	—
725	fresh, postripe	2-05-428	80.0	0.22	—	—	0.06	—	—	—	0.20	7.0	—	—	42.0	—	—
726			100.0	0.27	—	—	0.07	—	—	—	0.25	8.0	—	—	53.0	—	—
727	hay, sun-cured	1-05-418	93.0	0.31	—	0.15	0.20	1.85	—	—	0.22	15.0	—	165.0	34.0	0.37	30.0
728			100.0	0.33	—	0.16	0.21	2.00	—	—	0.24	16.0	—	178.0	36.0	0.40	32.0
	WHEY																	
729	dehy (Cattle)	4-01-182	93.0	0.86	0.07	0.13	0.76	1.15	0.65	1.04	0.11	47.0	—	169.0	6.0	—	3.0
730			100.0	0.92	0.08	0.14	0.82	1.23	0.70	1.12	0.12	50.0	—	181.0	6.0	—	3.0
731	low lactose, dehy	4-01-186	93.0	1.59	1.03	0.21	1.05	2.95	1.44	1.07	—	7.0	9.85	245.0	8.0	0.05	8.0
732	(Dried whey product) (Cattle)		100.0	1.71	1.10	0.23	1.12	3.16	1.54	1.15	—	8.0	10.55	262.0	9.0	0.06	8.0

TABLE 2   Composition of Important Feeds: Mineral Elements—*Continued*

Entry Number	Feed Name Description	International Feed Number	Dry Matter (%)	Calcium (%)	Chlorine (%)	Magnesium (%)	Phosphorus (%)	Potassium (%)	Sodium (%)	Sulfur (%)	Cobalt (mg/kg)	Copper (mg/kg)	Iodine (mg/kg)	Iron (mg/kg)	Manganese (mg/kg)	Selenium (mg/kg)	Zinc (mg/kg)
	WINTERFAT, COMMON *Eurotia lanata*																
733	fresh, stem-cured	2-26-142	80.0	1.58	—	—	0.09	—	—	—	—	—	—	—	—	—	—
734			100.0	1.98	—	—	0.12	—	—	—	—	—	—	—	—	—	—
	WOOD MOLASSES—SEE MOLASSES AND SYRUP																
	YEAST *Saccharomyces cerevisiae*																
735	brewers, dehy	7-05-527	93.0	0.12	0.07	0.25	1.40	1.67	0.07	0.42	0.18	33.0	0.36	109.0	6.0	0.91	39.0
736			100.0	0.13	0.08	0.27	1.49	1.79	0.08	0.45	0.20	35.0	0.38	117.0	6.0	0.98	41.0
737	irradiated, dehy	7-05-529	94.0	0.78	—	—	1.42	2.14	—	—	—	—	—	—	—	—	—
738			100.0	0.83	—	—	1.51	2.28	—	—	—	—	—	—	—	—	—
739	primary, dehy	7-05-533	93.0	0.36	0.02	0.36	1.72	—	—	0.57	—	—	—	300.0	4.0	—	—
740			100.0	0.39	0.02	0.39	1.86	—	—	0.62	—	—	—	324.0	4.0	—	—
	YEAST, TORULA *Torulopsis utilis*																
741	torula, dehy	7-05-534	93.0	0.50	0.02	0.17	1.59	1.90	0.04	0.55	0.03	13.0	2.51	118.0	8.0	1.00	93.0
742			100.0	0.54	0.02	0.18	1.71	2.04	0.04	0.59	0.03	14.0	2.69	126.0	9.0	1.08	100.0

TABLE 3   Composition of Important Feeds: Vitamins, Data Expressed As-Fed and Dry (100% Dry Matter)

Entry Number	Feed Name Description	International Feed Number	Dry Matter (%)	Carotene (Provitamin A) (mg/kg)	Vitamin A (IU/g)	Vitamin D₂ (IU/g)	Vitamin E (mg/kg)	Vitamin K (mg/kg)
				Fat-Soluble Vitamins				
	ALFALFA *Medicago sativa*							
001	fresh	2-00-196	24.0	45.0	—	46.0	—	—
002			100.0	185.0	—	191.0	—	—
003	hay, sun-cured	1-00-078	90.0	52.0	—	1,417.0	102.0	19.4
004			100.0	58.0	—	1,575.0	113.0	21.6
005	hay, sun-cured, early vegetative	1-00-050	90.0	181.0	—	1,806.0	—	—
006			100.0	201.0	—	2,007.0	—	—
007	hay, sun-cured, late vegetative	1-00-054	90.0	181.0	—	—	—	—
008			100.0	202.0	—	—	—	—
009	hay, sun-cured, early bloom	1-00-059	90.0	126.0	—	1,796.0	23.0	—
010			100.0	140.0	—	1,996.0	26.0	—
011	hay, sun-cured, full bloom	1-00-068	90.0	59.0	—	—	—	—
012			100.0	65.0	—	—	—	—
013	hay, sun-cured, mature	1-00-071	91.0	11.0	—	1,287.0	—	—
014			100.0	12.0	—	1,411.0	—	—
015	hay, sun-cured ground	1-00-111	91.0	122.0	—	—	60.0	7.8
016			100.0	135.0	—	—	66.0	8.6
017	leaves, sun-cured	1-00-146	89.0	79.0	—	333.0	—	—
018			100.0	88.0	—	373.0	—	—
019	meal dehy, 15% protein	1-00-022	90.0	74.0	—	—	82.0	9.6
020			100.0	82.0	—	—	91.0	10.6
021	meal dehy, 17% protein	1-00-023	92.0	120.0	—	—	111.0	8.2
022			100.0	131.0	—	—	121.0	9.0
023	meal dehy, 20% protein	1-00-024	92.0	159.0	—	—	151.0	14.2
024			100.0	174.0	—	—	165.0	15.5
025	meal dehy, 22% protein	1-07-851	93.0	235.0	—	—	221.0	11.6
026			100.0	253.0	—	—	238.0	12.6
027	silage	3-00-212	41.0	41.0	—	120.0	—	—
028			100.0	99.0	—	289.0	—	—
029	wilted silage	3-00-221	39.0	24.0	—	216.0	—	—
030			100.0	60.0	—	551.0	—	—
	BAHIAGRASS *Paspalum notatum*							
031	fresh	2-00-464	30.0	54.0	—	—	—	—
032			100.0	183.0	—	—	—	—
	BAKERY							
033	waste, dehy (Dried bakery	4-00-466	92.0	4.0	6.5	—	41.0	—
034	product)		100.0	5.0	7.0	—	45.0	—
	BARLEY *Hordeum vulgare*							
035	grain	4-00-549	88.0	2.0	—	—	22.0	0.2
036			100.0	2.0	—	—	25.0	0.2
037	grain, Pacific Coast	4-07-939	89.0	—	—	—	26.0	—
038			100.0	—	—	—	30.0	—
039	hay, sun-cured	1-00-495	87.0	46.0	—	963.0	—	—
040			100.0	53.0	—	1,103.0	—	—
041	malt sprouts, dehy	5-00-545	94.0	—	—	—	15.0	—
042			100.0	—	—	—	16.0	—
043	straw	1-00-498	91.0	2.0	—	603.0	—	—
044			100.0	2.0	—	662.0	—	—
	BEAN, NAVY *Phaseolus vulgaris*							
045	seeds	5-00-623	89.0	—	—	—	1.0	—
046			100.0	—	—	—	1.0	—

| Entry Number | Water Soluble Vitamins | | | | | | | | | |
	Biotin (mg/kg)	Choline (mg/kg)	Folic Acid (Folacin) (mg/kg)	Niacin (mg/kg)	Pantothenic Acid (mg/kg)	Riboflavin (mg/kg)	Thiamine (mg/kg)	Vitamin B_6 (mg/kg)	Vitamin B_{12} (μg/kg)	Xanthophylls (mg/kg)
001	0.12	378.0	0.6	12.0	8.5	3.3	1.4	1.6	—	—
002	0.49	1,556.0	2.5	49.0	34.9	13.4	5.9	6.7	—	—
003	0.18	—	3.2	38.0	25.7	12.1	2.7	—	—	33.0
004	0.20	—	3.6	42.0	28.6	13.4	3.0	—	—	37.0
005	—	—	—	—	—	—	—	—	—	—
006	—	—	—	—	—	—	—	—	—	—
007	—	—	—	—	—	—	—	—	—	—
008	—	—	—	—	—	—	—	—	—	—
009	—	—	—	—	—	—	—	—	—	—
010	—	—	—	—	—	—	—	—	—	—
011	—	—	—	—	—	—	—	—	—	—
012	—	—	—	—	—	—	—	—	—	—
013	—	—	—	—	—	—	—	—	—	—
014	—	—	—	—	—	—	—	—	—	—
015	0.29	1,162.0	3.7	38.0	26.3	12.5	3.8	4.0	—	112.0
016	0.33	1,283.0	4.1	42.0	29.1	13.8	4.2	4.4	—	123.0
017	0.28	1,062.0	5.8	47.0	29.0	20.6	4.6	—	—	—
018	0.31	1,189.0	6.5	53.0	32.4	23.1	5.2	—	—	—
019	0.25	1,573.0	1.6	42.0	20.7	10.6	3.0	6.3	—	171.0
020	0.28	1,739.0	1.7	46.0	22.9	11.7	3.3	6.9	—	189.0
021	0.33	1,370.0	4.4	37.0	29.8	12.9	3.4	7.1	—	263.0
022	0.36	1,494.0	4.8	40.0	32.4	14.1	3.7	7.7	—	287.0
023	0.35	1,418.0	3.0	48.0	35.5	15.2	5.4	8.8	—	282.0
024	0.39	1,547.0	3.2	52.0	38.8	16.6	5.9	9.6	—	308.0
025	0.33	1,605.0	5.1	50.0	39.0	17.6	5.9	8.3	—	330.0
026	0.36	1,729.0	5.5	54.0	42.9	19.0	6.3	8.9	—	356.0
027	—	—	—	—	—	—	—	—	—	—
028	—	—	—	—	—	—	—	—	—	—
029	—	—	—	—	—	—	—	—	—	—
030	—	—	—	—	—	—	—	—	—	—
031	—	—	—	—	—	—	—	—	—	—
032	—	—	—	—	—	—	—	—	—	—
033	0.07	923.0	0.2	26.0	8.3	1.4	2.9	4.3	—	2.0
034	0.07	1,005.0	0.2	28.0	9.0	1.5	3.2	4.7	—	2.0
035	0.15	1,039.0	0.6	83.0	8.1	1.6	4.4	6.4	—	—
036	0.17	1,177.0	0.6	94.0	9.1	1.8	5.0	7.3	—	—
037	0.15	982.0	0.5	47.0	7.1	1.5	4.2	2.9	—	—
038	0.17	1,102.0	0.6	53.0	8.0	1.7	4.7	3.3	—	—
039	—	—	—	—	—	—	—	—	—	—
040	—	—	—	—	—	—	—	—	—	—
041	4.12	1,606.0	0.2	51.0	8.9	3.0	8.4	9.5	—	—
042	4.40	1,713.0	0.2	54.0	9.5	3.2	8.9	10.2	—	—
043	—	—	—	—	—	—	—	—	—	—
044	—	—	—	—	—	—	—	—	—	—
045	0.11	1,341.0	1.3	25.0	2.1	1.8	6.3	0.3	—	—
046	0.12	1,499.0	1.4	28.0	2.3	2.0	7.1	0.3	—	—

TABLE 3 Composition of Important Feeds: Vitamins—*Continued*

Entry Number	Feed Name Description	International Feed Number	Dry Matter (%)	Fat-Soluble Vitamins				
				Carotene (Provitamin A) (mg/kg)	Vitamin A (IU/g)	Vitamin D$_2$ (IU/g)	Vitamin E (mg/kg)	Vitamin K (mg/kg)
	BEET, MANGELS *Beta vulgaris macrorhiza*							
047	roots, fresh	4-00-637	11.0	0.0	—	—	—	—
048			100.0	1.0	—	—	—	—
	BEET, SUGAR *Beta vulgaris altissima*							
	molasses—see Molasses and syrup							
049	pulp, dehy	4-00-669	91.0	0.0	—	577.0	—	—
050			100.0	0.0	—	637.0	—	—
	BERMUDAGRASS *Cynodon dactylon*							
051	fresh	2-00-712	34.0	104.0	—	—	—	—
052			100.0	310.0	—	—	—	—
053	hay, sun-cured	1-00-703	91.0	53.0	—	—	—	—
054			100.0	58.0	—	—	—	—
	BERMUDAGRASS, COASTAL *Cynodon dactylon*							
055	fresh	2-00-719	29.0	96.0	—	—	—	—
056			100.0	331.0	—	—	—	—
057	hay, sun-cured	1-00-716	90.0	95.0	—	—	—	—
058			100.0	105.0	—	—	—	—
	BIRDSFOOT TREFOIL—SEE TREFOIL, BIRDSFOOT							
	BLOOD							
059	meal	5-00-380	92.0	—	—	—	—	—
060			100.0	—	—	—	—	—
061	meal flash dehy	5-26-006	92.0	—	—	—	—	—
062			100.0	—	—	—	—	—
063	meal spray dehy (Blood flour)	5-00-381	93.0	—	—	—	—	—
064			100.0	—	—	—	—	—
	BLUEGRASS, CANADA *Poa compressa*							
065	fresh, early vegetative	2-00-763	26.0	104.0	—	—	—	—
066			100.0	400.0	—	—	—	—
067	hay, sun-cured	1-00-762	92.0	270.0	—	—	—	—
068			100.0	293.0	—	—	—	—
	BLUEGRASS, KENTUCKY *Poa pratensis*							
069	fresh	2-00-786	35.0	87.0	—	—	—	—
070			100.0	248.0	—	—	—	—
	BLUESTEM *Andropogon* spp							
071	fresh, early vegetative	2-00-821	27.0	59.0	—	—	—	—
072			100.0	219.0	—	—	—	—
	BREWERS							
073	grains, dehy	5-02-141	92.0	0.0	—	—	26.0	—
074			100.0	0.0	—	—	29.0	—

| Entry Num-ber | Water Soluble Vitamins | | | | | | | | | |
	Bio-tin (mg/kg)	Cho-line (mg/kg)	Folic Acid (Folacin) (mg/kg)	Niacin (mg/kg)	Pantothe-nic Acid (mg/kg)	Ribo-flavin (mg/kg)	Thia-mine (mg/kg)	Vita-min B_6 (mg/kg)	Vita-min B_{12} (μg/kg)	Xan-tho-phylls (mg/kg)
047	—	—	0.2	3.0	1.0	0.4	0.3	0.4	—	—
048	—	—	1.6	31.0	9.4	3.9	2.4	3.9	—	—
049	—	818.0	—	17.0	1.3	0.7	0.4	—	—	—
050	—	902.0	—	18.0	1.5	0.8	0.4	—	—	—
051	—	—	—	—	—	—	—	—	—	—
052	—	—	—	—	—	—	—	—	—	—
053	—	—	—	—	—	—	—	—	—	—
054	—	—	—	—	—	—	—	—	—	—
055	—	—	—	—	—	—	—	—	—	—
056	—	—	—	—	—	—	—	—	—	—
057	—	—	—	—	—	—	—	—	—	266.0
058	—	—	—	—	—	—	—	—	—	294.0
059	0.09	781.0	0.1	31.0	2.4	2.0	0.3	4.4	44.0	—
060	0.09	854.0	0.1	34.0	2.6	2.2	0.4	4.8	49.0	—
061	—	781.0	—	23.0	1.0	1.4	1.0	—	—	—
062	—	848.0	—	25.0	1.1	1.6	1.1	—	—	—
063	0.28	600.0	0.4	22.0	3.2	2.9	0.3	4.5	12.0	—
064	0.30	645.0	0.4	24.0	3.5	3.1	0.3	4.8	13.0	—
065	—	—	—	—	—	—	—	—	—	—
066	—	—	—	—	—	—	—	—	—	—
067	—	—	—	—	—	—	—	—	—	—
068	—	—	—	—	—	—	—	—	—	—
069	—	—	—	23.0	—	3.9	3.1	—	—	—
070	—	—	—	66.0	—	11.0	8.8	—	—	—
071	—	—	—	—	—	—	—	—	—	—
072	—	—	—	—	—	—	—	—	—	—
073	0.63	1,617.0	7.1	43.0	8.2	1.4	0.6	0.7	—	—
074	0.68	1,757.0	7.7	47.0	8.9	1.6	0.7	0.8	—	—

TABLE 3   Composition of Important Feeds: Vitamins—*Continued*

Entry Number	Feed Name Description	International Feed Number	Dry Matter (%)	Carotene (Provitamin A) (mg/kg)	Vitamin A (IU/g)	Vitamin D₂ (IU/g)	Vitamin E (mg/kg)	Vitamin K (mg/kg)
				Fat-Soluble Vitamins				
	BROME *Bromus* spp							
075	fresh	2-00-900	33.0	63.0	—	—	—	—
076			100.0	192.0	—	—	—	—
077	hay, sun-cured	1-00-890	91.0	31.0	—	948.0	—	—
078			100.0	34.0	—	1,047.0	—	—
	BROME, SMOOTH *Bromus inermis*							
079	fresh	2-00-963	32.0	102.0	—	32.0	—	—
080			100.0	315.0	—	99.0	—	—
081	hay, sun-cured	1-00-947	90.0	18.0	—	—	—	—
082			100.0	19.0	—	—	—	—
	BROOMCORN MILLET—SEE MILLET, PROSO							
	BUCKWHEAT, COMMON *Fagopyrum sagittatum*							
083	grain	4-00-994	88.0	—	—	—	—	—
084			100.0	—	—	—	—	—
	BUTTERMILK							
085	dehy (Cattle)	5-01-160	92.0	—	2.2	—	6.0	—
086			100.0	—	2.4	—	7.0	—
	CANARYGRASS, REED *Phalaris arundinacea*							
087	hay, sun-cured	1-01-104	91.0	23.0	—	—	—	—
088			100.0	26.0	—	—	—	—
	CARROT *Daucus* spp							
089	roots, fresh	4-01-145	12.0	80.0	—	—	7.0	—
090			100.0	678.0	—	—	60.0	—
	CASEIN							
091	dehy (cattle)	5-01-162	91.0	—	—	—	—	—
092			100.0	—	—	—	—	—
	CATTLE *Bos taurus*							
093	livers, fresh	5-01-166	28.0	—	122.7	—	7.0	—
094			100.0	—	439.1	—	25.0	—
095	lungs, fresh	5-07-941	21.0	—	0.7	—	3.0	—
096			100.0	—	3.3	—	13.0	—
	milk—see Milk skim milk—see Milk							
097	spleens, fresh	5-07-942	24.0	—	0.7	—	14.0	—
098			100.0	—	3.0	—	56.0	—
099	udders, fresh	5-07-943	20.0	—	1.8	—	10.0	—
100			100.0	—	9.0	—	49.0	—
	CEREALS							
101	screenings	4-02-156	90.0	—	—	—	—	—
102			100.0	—	—	—	—	—
103	screenings refuse	4-02-151	91.0	—	—	—	—	—
104			100.0	—	—	—	—	—
105	screenings uncleaned	4-02-153	92.0	—	—	—	—	—
106			100.0	—	—	—	—	—

Entry Number	Bio-tin (mg/kg)	Cho-line (mg/kg)	Folic Acid (Folacin) (mg/kg)	Niacin (mg/kg)	Pantothe-nic Acid (mg/kg)	Ribo-flavin (mg/kg)	Thia-mine (mg/kg)	Vita-min B$_6$ (mg/kg)	Vita-min B$_{12}$ (μg/kg)	Xan-tho-phylls (mg/kg)
	Water Soluble Vitamins									
075	—	—	—	—	—	2.5	1.0	—	—	—
076	—	—	—	—	—	7.7	3.1	—	—	—
077	—	—	—	—	—	—	—	—	—	—
078	—	—	—	—	—	—	—	—	—	—
079	—	—	—	—	—	2.5	1.0	—	—	—
080	—	—	—	—	—	7.7	3.1	—	—	—
081	—	612.0	—	—	—	—	—	—	—	—
082	—	677.0	—	—	—	—	—	—	—	—
083	—	439.0	—	18.0	11.5	4.7	3.7	—	—	—
084	—	501.0	—	21.0	13.1	5.4	4.2	—	—	—
085	0.29	1,746.0	0.4	9.0	37.0	30.6	3.4	2.4	20.0	—
086	0.31	1,891.0	0.4	9.0	40.1	33.1	3.7	2.6	21.0	—
087	—	—	—	—	—	8.1	3.5	—	—	—
088	—	—	—	—	—	8.9	3.9	—	—	—
089	0.01	—	0.1	7.0	3.6	0.6	0.7	1.4	—	—
090	0.07	—	1.2	58.0	30.1	4.9	5.8	12.0	—	—
091	0.04	208.0	0.5	1.0	2.7	1.5	0.4	0.4	—	—
092	0.05	229.0	0.5	1.0	2.9	1.7	0.5	0.5	—	—
093	0.98	1,424.0	2.3	75.0	46.1	25.8	1.8	5.0	426.0	—
094	3.51	5,093.0	8.4	269.0	164.9	92.2	6.3	18.0	1,523.0	—
095	0.03	1,693.0	0.2	11.0	0.5	1.8	0.6	0.4	90.0	—
096	0.12	7,933.0	0.9	49.0	2.6	8.4	2.8	1.8	423.0	—
097	0.04	491.0	1.2	6.0	2.0	3.7	0.7	0.3	60.0	—
098	0.16	2,036.0	4.8	25.0	8.2	15.3	3.1	1.3	247.0	—
099	0.06	877.0	0.1	21.0	9.5	3.0	6.6	1.4	114.0	—
100	0.30	4,320.0	0.3	102.0	46.7	14.6	32.7	6.8	562.0	—
101	—	1,044.0	1.1	10.0	12.8	1.8	—	—	—	—
102	—	1,163.0	1.2	12.0	14.3	2.0	—	—	—	—
103	—	—	—	47.0	23.0	0.7	0.5	2.4	—	—
104	—	—	—	52.0	25.3	0.7	0.5	2.6	—	—
105	—	—	—	72.0	21.2	0.6	0.6	2.2	—	—
106	—	—	—	79.0	23.1	0.7	0.6	2.4	—	—

TABLE 3   Composition of Important Feeds: Vitamins—*Continued*

Entry Number	Feed Name Description	International Feed Number	Dry Matter (%)	Carotene (Provitamin A) (mg/kg)	Vitamin A (IU/g)	Vitamin D₂ (IU/g)	Vitamin E (mg/kg)	Vitamin K (mg/kg)
	CHICKEN *Gallus domesticus*							
107	broilers, whole, fresh	5-07-945	24.0	—	7.3	—	—	—
108			100.0	—	30.0	—	—	—
109	feet, fresh	5-07-947	33.0	—	0.5	—	4.0	—
110			100.0	—	1.5	—	13.0	—
111	hens, whole, fresh	5-07-950	33.0	—	—	—	102.0	—
112			100.0	—	—	—	310.0	—
	CITRUS *Citrus* spp							
113	pulp fines (Dried citrus meal)	4-01-235	91.0	—	—	—	—	—
114			100.0	—	—	—	—	—
115	pulp without fines, dehy (Dried	4-01-237	91.0	0.0	—	—	—	—
116	citrus pulp)		100.0	0.0	—	—	—	—
	syrup—see Molasses and syrup							
	CLOVER, ALSIKE *Trifolium hybridum*							
117	fresh, early vegetative	2-01-314	19.0	73.0	—	—	—	—
118			100.0	385.0	—	—	—	—
119	hay, sun-cured	1-01-313	88.0	164.0	—	—	—	—
120			100.0	187.0	—	—	—	—
	CLOVER, CRIMSON *Trifolium incarnatum*							
121	fresh, early vegetative	2-20-890	18.0	43.0	—	—	—	—
122			100.0	238.0	—	—	—	—
123	hay, sun-cured	1-01-328	87.0	20.0	—	—	—	—
124			100.0	23.0	—	—	—	—
	CLOVER, LADINO *Trifolium repens*							
125	fresh, early vegetative	2-01-380	19.0	68.0	—	—	—	—
126			100.0	353.0	—	—	—	—
127	hay, sun-cured	1-01-378	90.0	75.0	—	—	—	—
128			100.0	83.0	—	—	—	—
	CLOVER, RED *Trifolium pratense*							
129	fresh	2-01-434	23.0	47.0	—	—	—	—
130			100.0	202.0	—	—	—	—
131	hay, sun-cured	1-01-415	89.0	18.0	—	1,694.0	—	—
132			100.0	20.0	—	1,914.0	—	—
	COCONUT *Cocos nucifera*							
133	meats, meal mech extd (Copra	5-01-572	92.0	—	—	—	—	—
134	meal)		100.0	—	—	—	—	—
135	meats, meal solv extd (Copra	5-01-573	91.0	—	—	—	—	—
136	meal)		100.0	—	—	—	—	—
	CORN, DENT YELLOW *Zea mays indentata*							
137	aerial part with ears, sun-cured	1-28-231	81.0	4.0	—	1,074.0	—	—
138	(Fodder)		100.0	4.0	—	1,323.0	—	—
139	cobs, ground	1-28-234	90.0	1.0	—	—	—	—
140			100.0	1.0	—	—	—	—
141	distillers grains, dehy	5-28-235	94.0	3.0	—	—	—	—
142			100.0	3.0	—	—	—	—

| Entry Num- ber | Water Soluble Vitamins | | | | | | | | | |
	Bio- tin (mg/kg)	Cho- line (mg/kg)	Folic Acid (Folacin) (mg/kg)	Niacin (mg/kg)	Pantothe- nic Acid (mg/kg)	Ribo- flavin (mg/kg)	Thia- mine (mg/kg)	Vita- min B_6 (mg/kg)	Vita- min B_{12} (μg/kg)	Xan- tho- phylls (mg/kg)
107	—	—	—	56.0	—	3.8	0.7	—	—	—
108	—	—	—	230.0	—	15.6	2.9	—	—	—
109	0.03	170.0	0.8	38.0	4.1	0.9	0.1	0.6	18.0	—
110	0.08	523.0	2.4	117.0	12.6	2.8	0.3	1.9	55.0	—
111	0.15	2,075.0	0.2	74.0	6.7	2.1	0.8	1.5	92.0	10.0
112	0.46	6,288.0	0.5	225.0	20.4	6.4	2.4	4.6	278.0	31.0
113	—	—	—	21.0	13.0	2.4	1.3	—	—	—
114	—	—	—	23.0	14.3	2.7	1.4	—	—	—
115	—	790.0	—	22.0	14.0	2.3	1.5	—	—	—
116	—	867.0	—	24.0	15.4	2.5	1.6	—	—	—
117	—	—	—	—	—	—	—	—	—	—
118	—	—	—	—	—	—	—	—	—	—
119	—	—	—	—	—	15.1	4.2	—	—	—
120	—	—	—	—	—	17.2	4.8	—	—	—
121	—	—	—	—	—	—	—	—	—	—
122	—	—	—	—	—	—	—	—	—	—
123	—	—	—	—	—	—	—	—	—	—
124	—	—	—	—	—	—	—	—	—	—
125	—	—	—	—	—	—	—	—	—	—
126	—	—	—	—	—	—	—	—	—	—
127	—	—	—	10.0	1.0	15.2	3.8	—	—	—
128	—	—	—	11.0	1.1	17.0	4.2	—	—	—
129	—	—	—	19.0	—	4.5	1.5	—	—	—
130	—	—	—	81.0	—	19.2	6.6	—	—	—
131	0.09	—	—	38.0	9.9	15.7	2.0	—	—	—
132	0.11	—	—	43.0	11.2	17.8	2.2	—	—	—
133	—	956.0	1.4	24.0	6.3	3.2	0.8	—	—	—
134	—	1,036.0	1.5	26.0	6.8	3.4	0.8	—	—	—
135	—	1,083.0	0.3	26.0	6.3	3.3	0.6	4.4	—	—
136	—	1,189.0	0.3	28.0	6.9	3.7	0.7	4.8	—	—
137	—	—	—	—	—	—	—	—	—	—
138	—	—	—	—	—	—	—	—	—	—
139	—	—	—	7.0	3.8	1.0	0.9	—	—	—
140	—	—	—	8.0	4.2	1.1	1.0	—	—	—
141	0.49	1,180.0	0.9	37.0	11.7	5.2	1.7	4.4	—	2.0
142	0.52	1,262.0	0.9	40.0	12.5	5.6	1.8	4.7	—	2.0

TABLE 3 Composition of Important Feeds: Vitamins—*Continued*

Entry Number	Feed Name Description	International Feed Number	Dry Matter (%)	Fat-Soluble Vitamins				
				Carotene (Provita-min A) (mg/kg)	Vita-min A (IU/g)	Vita-min D₂ (IU/g)	Vita-min E (mg/kg)	Vita-min K (mg/kg)
143	distillers grains with solubles,	5-28-236	92.0	3.0	—	551.0	40.0	—
144	dehy		100.0	3.0	—	600.0	43.0	—
145	distillers solubles, dehy	5-28-237	93.0	1.0	—	—	46.0	—
146			100.0	1.0	—	—	49.0	—
147	ears, ground (Corn and cob	4-28-238	87.0	3.0	—	—	18.0	—
148	meal)		100.0	4.0	—	—	20.0	—
149	germs, meal wet milled solv extd	5-28-240	91.0	2.0	—	—	85.0	—
150			100.0	2.0	—	—	94.0	—
151	gluten, meal	5-28-241	91.0	16.0	—	—	31.0	—
152			100.0	18.0	—	—	34.0	—
153	gluten, meal, 60% protein	5-28-242	90.0	30.0	—	—	24.0	—
154			100.0	34.0	—	—	26.0	—
155	gluten with bran (Corn gluten	5-28-243	90.0	6.0	—	—	12.0	—
156	feed)		100.0	7.0	—	—	14.0	—
157	grain	4-02-935	89.0	2.0	—	—	22.0	0.2
158			100.0	3.0	—	—	25.0	0.2
159	grain, opaque 2 (High lysine)	4-28-253	90.0	5.0	—	—	—	—
160			100.0	5.0	—	—	—	—
161	grits (Hominy grits)	4-03-010	88.0	3.0	—	—	—	—
162			100.0	3.0	—	—	—	—
163	grits by-product (Hominy feed)	4-03-011	90.0	9.0	—	—	—	—
164			100.0	10.0	—	—	—	—
165	silage	3-02-912	30.0	13.0	—	132.0	—	—
166			100.0	43.0	—	439.0	—	—
167	silage, aerial part without ears	3-28-251	31.0	5.0	—	—	—	—
168	without husks (Stalklage) (Stover)		100.0	15.0	—	—	—	—
169	silage, well-eared	3-28-250	33.0	15.0	—	40.0	—	—
170			100.0	45.0	—	119.0	—	—
	CORN, SWEET *Zea mays saccharata*							
171	process residue, fresh	2-02-975	77.0	10.0	—	—	—	—
172			100.0	14.0	—	—	—	—
173	process residue, silage	3-07-955	32.0	4.0	—	—	—	—
174			100.0	13.0	—	—	—	—
	COTTON *Gossypium* spp							
175	seeds, meal mech extd, 41%	5-01-617	93.0	0.0	—	—	32.0	—
176	protein		100.0	0.0	—	—	35.0	—
177	seeds, meal prepressed solv extd,	5-07-872	91.0	—	—	—	—	—
178	41% protein		100.0	—	—	—	—	—
179	seeds, meal prepressed solv extd,	5-07-873	91.0	—	—	—	—	—
180	44% protein		100.0	—	—	—	—	—
181	seeds, meal solv extd, 41%	5-01-621	91.0	—	—	—	16.0	—
182	protein		100.0	—	—	—	17.0	—
183	seeds without hulls, meal pre-	5-07-874	93.0	—	—	—	11.0	—
184	pressed solv extd, 50% protein		100.0	—	—	—	12.0	—
	COWPEA, COMMON *Vigna sinensis*							
185	hay, sun-cured	1-01-645	90.0	31.0	—	—	—	—
186			100.0	35.0	—	—	—	—
	CRAB *Callinectes sapidus– Cancer* spp							
187	process residue, meal (Crab meal)	5-01-663	92.0	—	—	—	—	—
188			100.0	—	—	—	—	—

Entry Number	Water Soluble Vitamins									
	Biotin (mg/kg)	Choline (mg/kg)	Folic Acid (Folacin) (mg/kg)	Niacin (mg/kg)	Pantothenic Acid (mg/kg)	Riboflavin (mg/kg)	Thiamine (mg/kg)	Vitamin B_6 (mg/kg)	Vitamin B_{12} (μg/kg)	Xanthophylls (mg/kg)
143	0.78	2,574.0	0.9	73.0	14.0	9.1	2.9	5.0	—	10.0
144	0.85	2,803.0	1.0	79.0	15.3	10.0	3.1	5.4	—	10.0
145	1.66	4,778.0	1.3	124.0	23.3	21.1	6.7	8.8	3.0	2.0
146	1.79	5,151.0	1.4	134.0	25.2	22.7	7.3	9.5	3.0	2.0
147	0.03	357.0	0.2	17.0	4.2	0.9	2.9	6.0	—	11.0
148	0.04	412.0	0.3	20.0	4.8	1.0	3.3	6.9	—	13.0
149	0.22	1,627.0	0.2	30.0	4.2	3.9	4.5	6.2	—	—
150	0.24	1,785.0	0.2	33.0	4.6	4.2	4.9	6.8	—	—
151	0.18	357.0	0.3	51.0	10.2	1.6	0.2	8.0	—	175.0
152	0.20	391.0	0.3	55.0	11.2	1.8	0.2	8.8	—	191.0
153	0.19	352.0	0.3	60.0	3.5	2.0	0.2	6.9	—	294.0
154	0.21	390.0	0.3	66.0	3.9	2.2	0.3	7.6	—	326.0
155	0.33	1,515.0	0.3	71.0	13.6	2.2	2.0	13.3	—	37.0
156	0.36	1,684.0	0.3	79.0	15.1	2.5	2.2	14.8	—	42.0
157	0.07	502.0	0.3	25.0	5.9	1.2	3.4	4.7	—	17.0
158	0.08	567.0	0.3	28.0	6.6	1.4	3.8	5.3	—	19.0
159	—	518.0	—	19.0	4.7	1.1	—	—	—	—
160	—	575.0	—	22.0	5.2	1.2	—	—	—	—
161	0.02	—	—	11.0	2.2	0.6	1.4	2.6	—	—
162	0.03	—	—	12.0	2.5	0.7	1.6	3.0	—	—
163	0.13	1,155.0	0.3	47.0	8.2	2.1	8.1	11.0	—	4.0
164	0.15	1,280.0	0.3	52.0	9.1	2.3	8.9	12.1	—	4.0
165	—	—	—	14.0	—	—	—	—	—	—
166	—	—	—	47.0	—	—	—	—	—	—
167	—	—	—	—	—	—	—	—	—	—
168	—	—	—	—	—	—	—	—	—	—
169	—	—	—	—	—	—	—	—	—	—
170	—	—	—	—	—	—	—	—	—	—
171	—	—	—	—	—	—	—	—	—	—
172	—	—	—	—	—	—	—	—	—	—
173	—	—	—	—	—	—	—	—	—	—
174	—	—	—	—	—	—	—	—	—	—
175	1.11	2,747.0	2.1	35.0	10.4	5.2	6.5	5.0	—	—
176	1.19	2,965.0	2.3	38.0	11.2	5.7	7.0	5.4	—	—
177	0.56	2,844.0	2.5	40.0	7.4	4.4	3.3	4.2	—	—
178	0.61	3,141.0	2.8	44.0	8.2	4.9	3.7	4.6	—	—
179	—	2,685.0	0.9	46.0	14.5	4.7	—	—	—	—
180	—	2,939.0	1.0	51.0	15.9	5.1	—	—	—	—
181	0.97	2,787.0	1.4	41.0	13.7	4.8	6.6	5.6	—	—
182	1.06	3,056.0	1.5	45.0	15.0	5.2	7.3	6.2	—	—
183	0.44	2,962.0	0.9	45.0	14.3	4.9	8.2	6.3	—	—
184	0.48	3,184.0	1.0	48.0	15.4	5.3	8.8	6.8	—	—
185	—	—	—	—	—	—	—	—	—	—
186	—	—	—	—	—	—	—	—	—	—
187	0.07	2,011.0	0.1	45.0	6.5	6.1	0.4	6.6	438.0	—
188	0.07	2,179.0	0.1	49.0	7.0	6.7	0.5	7.2	475.0	—

TABLE 3   Composition of Important Feeds: Vitamins—*Continued*

Entry Number	Feed Name Description	International Feed Number	Dry Matter (%)	Carotene (Provitamin A) (mg/kg)	Vitamin A (IU/g)	Vitamin D₂ (IU/g)	Vitamin E (mg/kg)	Vitamin K (mg/kg)
				Fat-Soluble Vitamins				
	DISTILLERS GRAINS—SEE CORN, SEE SORGHUM							
	FESCUE, MEADOW *Festuca elatior*							
189	fresh	2-01-920	28.0	104.0	—	—	40.0	—
190			100.0	376.0	—	—	143.0	—
191	hay, sun-cured	1-01-912	88.0	64.0	—	—	119.0	—
192			100.0	73.0	—	—	136.0	—
	FISH							
193	solubles, condensed	5-01-969	50.0	1.0	—	—	—	—
194			100.0	3.0	—	—	—	—
195	solubles, dehy	5-01-971	93.0	—	—	—	6.0	—
196			100.0	—	—	—	7.0	—
	FISH, ALEWIFE *Pomolobus pseudoharengus*							
197	meal mech extd	5-09-830	90.0	—	3.9	—	—	—
198			100.0	—	4.3	—	—	—
	FISH, ANCHOVY *Engraulis ringen*							
199	meal mech extd	5-01-985	92.0	—	—	—	5.0	—
200			100.0	—	—	—	5.0	—
	FISH, HERRING *Clupea harengus*							
201	meal mech extd	5-02-000	92.0	—	—	—	22.0	2.2
202			100.0	—	—	—	24.0	2.3
	FISH, MACKEREL, ATLANTIC *Scomber scombrus*							
203	whole, fresh	5-07-971	30.0	—	25.8	—	10.0	—
204			100.0	—	85.5	—	34.0	—
	FISH, MENHADEN *Brevoortia tyrannus*							
205	meal mech extd	5-02-009	92.0	—	—	—	12.0	—
206			100.0	—	—	—	13.0	—
	FISH, REDFISH *Sciaenops ocellata*							
207	meal mech extd	5-07-973	93.0	—	—	—	6.0	—
208			100.0	—	—	—	6.0	—
	FISH, SALMON *Oncorhynchus* spp–*Salmo* spp							
209	meal mech extd	5-02-012	93.0	—	—	—	—	—
210			100.0	—	—	—	—	—
	FISH, SARDINE *Clupea* spp–*Sardinops* spp							
211	meal mech extd	5-02-015	93.0	—	—	—	—	—
212			100.0	—	—	—	—	—

| Entry Number | Water Soluble Vitamins | | | | | | | | | |
	Bio-tin (mg/kg)	Cho-line (mg/kg)	Folic Acid (Folacin) (mg/kg)	Niacin (mg/kg)	Pantothe-nic Acid (mg/kg)	Ribo-flavin (mg/kg)	Thia-mine (mg/kg)	Vita-min B_6 (mg/kg)	Vita-min B_{12} (μg/kg)	Xan-tho-phylls (mg/kg)
189	—	—	—	—	—	3.3	4.6	—	—	—
190	—	—	—	—	—	12.0	16.8	—	—	—
191	—	—	—	—	—	—	—	—	—	—
192	—	—	—	—	—	—	—	—	—	—
193	0.14	3,389.0	0.2	175.0	35.5	12.6	5.0	12.1	505.0	—
194	0.28	6,759.0	0.4	350.0	70.8	25.2	10.0	24.2	1,007.0	—
195	0.39	5,518.0	0.6	255.0	50.3	13.5	7.4	24.0	485.0	—
196	0.43	5,954.0	0.6	276.0	54.3	14.6	8.0	25.9	524.0	—
197	—	4,636.0	—	30.0	8.2	2.8	0.1	—	311.0	—
198	—	5,160.0	—	33.0	9.1	3.2	0.1	—	346.0	—
199	0.20	3,709.0	0.2	82.0	10.0	7.5	0.5	4.6	214.0	—
200	0.21	4,036.0	0.2	89.0	10.9	8.2	0.5	5.0	233.0	—
201	0.48	5,286.0	0.3	85.0	16.8	10.1	0.4	4.8	429.0	—
202	0.52	5,752.0	0.4	93.0	18.2	11.0	0.4	5.2	467.0	—
203	0.04	1,035.0	2.6	7.0	5.4	2.9	0.9	0.4	228.0	—
204	0.12	3,422.0	8.5	24.0	17.9	9.6	2.9	1.2	753.0	—
205	0.18	3,112.0	0.2	55.0	8.6	4.8	0.6	4.7	122.0	—
206	0.20	3,398.0	0.2	60.0	9.4	5.2	0.6	5.1	133.0	—
207	0.17	3,429.0	—	41.0	8.4	7.0	0.2	—	141.0	—
208	0.18	3,681.0	—	44.0	9.0	7.5	0.2	—	152.0	—
209	—	2,783.0	—	25.0	6.9	5.8	0.9	—	—	—
210	—	2,990.0	—	27.0	7.4	6.2	0.9	—	—	—
211	0.10	3,277.0	—	75.0	11.0	5.4	0.3	—	238.0	—
212	0.11	3,518.0	—	81.0	11.8	5.8	0.3	—	256.0	—

TABLE 3   Composition of Important Feeds: Vitamins—*Continued*

Entry Number	Feed Name Description	International Feed Number	Dry Matter (%)	Fat-Soluble Vitamins				
				Carotene (Provitamin A) (mg/kg)	Vitamin A (IU/g)	Vitamin D$_2$ (IU/g)	Vitamin E (mg/kg)	Vitamin K (mg/kg)
	FISH, TUNA *Thunnus thynnus–Thunnus albacares*							
213	meal mech extd	5-02-023	93.0	—	—	—	6.0	—
214			100.0	—	—	—	6.0	—
	FISH, WHITE *Gadidae* (family)–*Lophiidae* (family)							
215	meal mech extd	5-02-025	91.0	—	—	—	9.0	—
216			100.0	—	—	—	10.0	—
	FLAX *Linum usitatissimum*							
217	seeds, meal mech extd (Linseed	5-02-045	91.0	0.0	—	—	8.0	—
218	meal)		100.0	0.0	—	—	9.0	—
219	seeds, meal solv extd (Linseed	5-02-048	90.0	—	—	—	14.0	—
220	meal)		100.0	—	—	—	15.0	—
	GALLETA *Hilaria jamesii*							
221	fresh, stem-cured	2-05-594	71.0	0.0	—	—	—	—
222			100.0	0.0	—	—	—	—
	GRAMA *Bouteloua* spp							
223	fresh, mature	2-02-166	63.0	19.0	—	—	—	—
224			100.0	30.0	—	—	—	—
	GRAPE *Vitis* spp							
225	marc, dehy (Pomace)	1-02-208	91.0	—	—	—	—	—
226			100.0	—	—	—	—	—
	GROUNDNUT—SEE PEANUT							
	HOG MILLET—SEE MILLET, PROSO							
	HOMINY FEED—SEE CORN							
	HORSE *Equus caballus*							
227	meat, fresh	5-07-980	29.0	—	0.6	—	7.0	—
228			100.0	—	2.0	—	25.0	—
	JOHNSONGRASS—SEE SORGHUM, JOHNSONGRASS							
	KENTUCKY BLUEGRASS—SEE BLUEGRASS, KENTUCKY							
	LESPEDEZA, COMMON–LESPEDEZA, KOREAN *Lespedeza striata–Lespedeza stipulacea*							
229	hay, sun-cured, late vegetative	1-260-024	92.0	133.0	—	—	—	—
230			100.0	145.0	—	—	—	—
231	hay, sun-cured, early bloom	1-26-025	93.0	128.0	—	—	—	—
232			100.0	138.0	—	—	—	—
233	hay, sun-cured, midbloom	1-26-026	93.0	51.0	—	—	—	—
234			100.0	55.0	—	—	—	—
235	hay, sun-cured, full bloom	1-26-027	93.0	12.0	—	—	—	—
236			100.0	13.0	—	—	—	—

| Entry Number | Water Soluble Vitamins | | | | | | | | | |
	Biotin (mg/kg)	Choline (mg/kg)	Folic Acid (Folacin) (mg/kg)	Niacin (mg/kg)	Pantothenic Acid (mg/kg)	Riboflavin (mg/kg)	Thiamine (mg/kg)	Vitamin B6 (mg/kg)	Vitamin B12 (µg/kg)	Xanthophylls (mg/kg)
213	0.20	2,994.0	—	144.0	7.7	6.8	1.5	—	300.0	—
214	0.22	3,227.0	—	155.0	8.4	7.3	1.6	—	324.0	—
215	0.08	3,099.0	0.3	59.0	9.9	9.1	1.7	5.9	90.0	—
216	0.09	3,397.0	0.4	65.0	10.9	10.0	1.8	6.5	98.0	—
217	0.33	1,780.0	2.8	37.0	14.3	3.2	4.2	5.5	—	2.0
218	0.36	1,962.0	3.1	41.0	15.8	3.5	4.6	6.1	—	2.0
219	—	1,393.0	1.3	33.0	14.7	2.9	7.5	—	—	—
220	—	1,544.0	1.4	37.0	16.3	3.2	8.4	—	—	—
221	—	—	—	—	—	—	—	—	—	—
222	—	—	—	—	—	—	—	—	—	—
223	—	—	—	—	—	—	—	—	—	—
224	—	—	—	—	—	—	—	—	—	—
225	—	254.0	—	18.0	3.1	2.2	—	—	—	—
226	—	279.0	—	20.0	3.4	2.5	—	—	—	—
227	0.02	304.0	0.2	5.0	1.4	—	0.4	0.2	41.0	—
228	0.08	1,043.0	0.8	16.0	4.8	—	1.4	0.7	142.0	—
229	—	—	—	—	—	—	—	—	—	—
230	—	—	—	—	—	—	—	—	—	—
231	—	—	—	—	—	—	—	—	—	—
232	—	—	—	—	—	—	—	—	—	—
233	—	—	—	—	—	—	—	—	—	—
234	—	—	—	—	—	—	—	—	—	—
235	—	—	—	—	—	—	—	—	—	—
236	—	—	—	—	—	—	—	—	—	—

TABLE 3   Composition of Important Feeds: Vitamins—*Continued*

Entry Number	Feed Name Description	International Feed Number	Dry Matter (%)	Fat-Soluble Vitamins Carotene (Provitamin A) (mg/kg)	Vitamin A (IU/g)	Vitamin D₂ (IU/g)	Vitamin E (mg/kg)	Vitamin K (mg/kg)
	LESPEDEZA, CHINESE *Lespedeza cuneata*							
237	hay, sun-cured, late vegetative	1-09-172	92.0	37.0	—	—	—	—
238			100.0	40.0	—	—	—	—
	LINSEED—SEE FLAX							
	LIVERS							
239	meal	5-00-389	92.0	—	—	—	—	—
240			100.0	—	—	—	—	—
	MAIZE—SEE CORN							
	MANGELS—SEE BEET							
	MEAT							
241	meal rendered	5-00-385	94.0	—	—	—	1.0	—
242			100.0	—	—	—	1.0	—
243	with blood, meal rendered	5-00-386	92.0	—	—	—	—	—
244	(Tankage)		100.0	—	—	—	—	—
245	with bone, meal rendered	5-00-388	93.0	—	—	—	1.0	—
246			100.0	—	—	—	1.0	—
	MILK							
247	dehy (Cattle)	5-01-167	96.0	—	11.7	338.0	—	—
248			100.0	—	12.3	353.0	—	—
249	skimmed dehy (Cattle)	5-01-175	94.0	—	—	419.0	9.0	—
250			100.0	—	—	446.0	10.0	—
	MILLET, PEARL—SEE PEARLMILLET							
	MILLET, PROSO *Panicum miliaceum*							
251	grain	4-03-120	90.0	—	—	—	—	—
252			100.0	—	—	—	—	—
	MOLASSES AND SYRUP							
253	beet, sugar, molasses, more than	4-00-668	78.0	—	—	—	4.0	—
254	48% invert sugar more than 79.5 degrees brix		100.0	—	—	—	5.0	—
255	citrus, syrup (Citrus molasses)	4-01-241	68.0	—	—	—	—	—
256			100.0	—	—	—	—	—
257	sugarcane, molasses, more than	4-04-696	75.0	—	—	—	5.0	—
258	46% invert sugar more than 79.5 degrees brix (Black strap)		100.0	—	—	—	7.0	—
	OATS *Avena sativa*							
259	breakfast cereal by-product, less	4-03-303	91.0	—	—	—	24.0	—
260	than 4% fiber (Feeding oat meal) (Oat middlings)		100.0	—	—	—	26.0	—
261	grain	4-03-309	89.0	—	—	—	14.0	—
262			100.0	—	—	—	15.0	—
263	grain, Pacific Coast	4-07-999	91.0	—	—	—	20.0	—
264			100.0	—	—	—	22.0	—
265	groats	4-03-331	90.0	—	—	—	15.0	—
266			100.0	—	—	—	16.0	—

| Entry Number | Water Soluble Vitamins | | | | | | | | | |
	Biotin (mg/kg)	Choline (mg/kg)	Folic Acid (Folacin) (mg/kg)	Niacin (mg/kg)	Pantothenic Acid (mg/kg)	Riboflavin (mg/kg)	Thiamine (mg/kg)	Vitamin B$_6$ (mg/kg)	Vitamin B$_{12}$ (μg/kg)	Xanthophylls (mg/kg)
237	—	—	—	—	—	—	—	—	—	—
238	—	—	—	—	—	—	—	—	—	—
239	0.02	11,359.0	5.6	205.0	29.1	36.2	0.2	—	501.0	—
240	0.02	12,281.0	6.0	221.0	31.5	39.1	0.2	—	542.0	—
241	0.12	2,041.0	0.4	56.0	6.1	5.2	0.2	2.7	67.0	—
242	0.13	2,177.0	0.4	60.0	6.5	5.6	0.2	2.9	72.0	—
243	—	2,201.0	1.5	37.0	2.6	2.2	0.3	—	135.0	—
244	—	2,391.0	1.7	40.0	2.8	2.4	0.4	—	147.0	—
245	0.10	2,044.0	0.4	49.0	4.1	4.5	0.2	8.7	108.0	—
246	0.11	2,196.0	0.4	53.0	4.4	4.9	0.2	9.4	116.0	—
247	0.38	—	—	8.0	22.8	19.7	3.8	4.7	—	—
248	0.40	—	—	9.0	23.8	20.6	3.9	4.9	—	—
249	0.33	1,390.0	0.6	11.0	36.3	19.3	3.7	4.2	51.0	—
250	0.35	1,480.0	0.7	12.0	38.6	20.5	3.9	4.5	54.0	—
251	—	440.0	—	23.0	11.0	3.8	7.3	—	—	—
252	—	489.0	—	26.0	12.2	4.2	8.1	—	—	—
253	—	826.0	—	41.0	4.5	2.3	—	—	—	—
254	—	1,063.0	—	53.0	5.8	2.9	—	—	—	—
255	—	—	—	27.0	12.7	6.2	—	—	—	—
256	—	—	—	40.0	18.8	9.2	—	—	—	—
257	0.69	754.0	0.1	37.0	37.5	2.8	0.9	4.2	—	—
258	0.92	1,012.0	0.1	49.0	50.3	3.8	1.2	5.7	—	—
259	0.22	1,149.0	0.5	22.0	16.9	1.7	7.0	—	—	—
260	0.24	1,267.0	0.5	25.0	18.6	1.9	7.7	—	—	—
261	0.28	992.0	0.4	14.0	7.8	1.5	6.3	2.5	—	—
262	0.31	1,116.0	0.4	16.0	8.8	1.7	7.1	2.8	—	—
263	—	917.0	—	14.0	11.7	1.2	—	—	—	—
264	—	1,009.0	—	16.0	12.8	1.3	—	—	—	—
265	—	1,132.0	0.5	10.0	13.8	1.2	6.5	1.1	—	—
266	—	1,264.0	0.6	11.0	15.4	1.3	7.2	1.2	—	—

TABLE 3 Composition of Important Feeds: Vitamins—*Continued*

Entry Number	Feed Name Description	International Feed Number	Dry Matter (%)	Carotene (Provitamin A) (mg/kg)	Vitamin A (IU/g)	Vitamin D₂ (IU/g)	Vitamin E (mg/kg)	Vitamin K (mg/kg)
267	hay, sun-cured	1-03-280	91.0	25.0	—	1,410.0	—	—
268			100.0	28.0	—	1,544.0	—	—
269	hulls	1-03-281	92.0	—	—	—	—	—
270			100.0	—	—	—	—	—
271	silage	3-03-298	31.0	14.0	—	—	—	—
272			100.0	45.0	—	—	—	—
273	straw	1-03-283	92.0	4.0	—	609.0	—	—
274			100.0	4.0	—	662.0	—	—
	ORCHARDGRASS *Dactylis glomerata*							
275	fresh	2-03-451	27.0	81.0	—	—	117.0	—
276			100.0	302.0	—	—	436.0	—
277	hay, sun-cured	1-03-438	91.0	20.0	—	—	174.0	—
278			100.0	22.0	—	—	191.0	—
	PANGOLAGRASS *Digitaria decumbens*							
279	fresh	2-03-493	21.0	13.0	—	—	—	—
280			100.0	62.0	—	—	—	—
	PEA *Pisum* spp							
281	seeds	5-03-600	89.0	1.0	—	—	3.0	—
282			100.0	1.0	—	—	3.0	—
283	vines without seeds, silage	3-03-596	25.0	46.0	—	—	—	—
284			100.0	189.0	—	—	—	—
	PEANUT *Arachis hypogaea*							
285	hay, sun-cured	1-03-619	91.0	32.0	—	3,273.0	—	—
286			100.0	35.0	—	3,601.0	—	—
287	kernels, meal mech extd (Peanut	5-03-649	93.0	0.0	—	—	2.0	—
288	meal)		100.0	0.0	—	—	3.0	—
289	kernels, meal solv extd (Peanut	5-03-650	92.0	—	—	—	—	—
290	meal)		100.0	—	—	—	—	—
	PEARLMILLET *Pennisetum glaucum*							
291	fresh	2-03-115	21.0	38.0	—	—	—	—
292			100.0	183.0	—	—	—	—
293	silage	3-20-903	30.0	7.0	—	—	—	—
294			100.0	25.0	—	—	—	—
	PINEAPPLE *Ananas comosus*							
295	process residue, dehy (Pineapple	4-03-722	87.0	47.0	—	—	—	—
296	bran)		100.0	54.0	—	—	—	—
	POTATO *Solanum tuberosum*							
297	tubers, dehy	4-07-850	91.0	—	—	—	—	—
298			100.0	—	—	—	—	—
	POULTRY							
299	by-product, meal rendered	5-03-798	93.0	—	—	—	2.0	—
300	(Viscera with feet with heads)		100.0	—	—	—	2.0	—
301	feathers, hydrolyzed	5-03-795	93.0	—	—	—	—	—
302			100.0	—	—	—	—	—

Entry Number	Water Soluble Vitamins									
	Bio-tin (mg/kg)	Cho-line (mg/kg)	Folic Acid (Folacin) (mg/kg)	Niacin (mg/kg)	Pantothe-nic Acid (mg/kg)	Ribo-flavin (mg/kg)	Thia-mine (mg/kg)	Vita-min B_6 (mg/kg)	Vita-min B_{12} (μg/kg)	Xan-tho-phylls (mg/kg)
267	—	—	—	—	—	4.8	3.0	—	—	—
268	—	—	—	—	—	5.3	3.3	—	—	—
269	—	260.0	1.0	9.0	3.1	1.7	0.6	2.2	—	—
270	—	281.0	1.0	10.0	3.4	1.9	0.7	2.4	—	—
271	—	304.0	—	—	—	—	—	—	—	—
272	—	992.0	—	—	—	—	—	—	—	—
273	—	215.0	—	—	—	—	—	—	—	—
274	—	234.0	—	—	—	—	—	—	—	—
275	—	—	—	—	—	—	2.0	—	—	—
276	—	—	—	—	—	—	7.3	—	—	—
277	—	—	—	—	—	6.2	2.6	—	—	—
278	—	—	—	—	—	6.8	2.9	—	—	—
279	—	—	—	—	—	—	—	—	—	—
280	—	—	—	—	—	—	—	—	—	—
281	0.20	589.0	0.3	32.0	18.7	1.8	4.6	1.5	—	—
282	0.22	662.0	0.3	36.0	21.0	2.0	5.2	1.7	—	—
283	—	—	—	—	—	—	—	—	—	—
284	—	—	—	—	—	—	—	—	—	—
285	—	—	—	—	—	8.8	—	—	—	—
286	—	—	—	—	—	9.7	—	—	—	—
287	0.33	1,900.0	0.7	172.0	46.0	8.1	6.1	7.4	—	—
288	0.35	2,052.0	0.7	186.0	49.7	8.8	6.6	8.0	—	—
289	0.33	1,951.0	0.7	173.0	46.6	9.1	5.7	6.4	—	—
290	0.36	2,120.0	0.7	188.0	50.7	9.8	6.2	6.9	—	—
291	—	—	—	—	—	—	—	—	—	—
292	—	—	—	—	—	—	—	—	—	—
293	—	—	—	—	—	—	—	—	—	—
294	—	—	—	—	—	—	—	—	—	—
295	—	—	—	—	—	—	—	—	—	—
296	—	—	—	—	—	—	—	—	—	—
297	0.10	2,626.0	0.6	33.0	20.1	1.0	—	14.1	—	—
298	0.11	2,879.0	0.7	37.0	22.0	1.1	—	15.5	—	—
299	0.09	6,029.0	0.5	47.0	11.1	10.5	0.2	4.4	301.0	—
300	0.09	6,451.0	0.5	50.0	11.8	11.2	0.2	4.7	322.0	—
301	0.04	895.0	0.2	21.0	9.0	2.0	0.1	3.0	83.0	—
302	0.05	962.0	0.2	23.0	9.7	2.1	0.1	3.2	90.0	—

TABLE 3 Composition of Important Feeds: Vitamins—*Continued*

Entry Number	Feed Name Description	International Feed Number	Dry Matter (%)	Carotene (Provitamin A) (mg/kg)	Vitamin A (IU/g)	Vitamin D₂ (IU/g)	Vitamin E (mg/kg)	Vitamin K (mg/kg)
				Fat-Soluble Vitamins				
	PRAIRIE PLANTS, MIDWEST							
303	hay, sun-cured	1-03-191	92.0	22.0	—	1,158.0	—	—
304			100.0	24.0	—	1,264.0	—	—
	RAPE *Brassica* spp							
305	fresh, early vegetative	2-03-865	18.0	28.0	—	—	—	—
306			100.0	155.0	—	—	—	—
307	seeds, meal mech extd	5-03-870	92.0	—	—	—	19.0	—
308			100.0	—	—	—	20.0	—
309	seeds, meal solv extd	5-03-871	91.0	—	—	—	—	—
310			100.0	—	—	—	—	—
	REDTOP *Agrostis alba*							
311	fresh	2-03-897	29.0	64.0	—	—	—	—
312			100.0	217.0	—	—	—	—
	RICE *Oryza sativa*							
313	bran with germ (Rice bran)	4-03-928	91.0	—	—	—	60.0	—
314			100.0	—	—	—	66.0	—
315	grain, ground (Ground rough	4-03-938	89.0	—	—	—	10.0	—
316	rice) (Ground paddy rice)		100.0	—	—	—	11.0	—
317	grain, polished and broken	4-03-932	89.0	—	—	—	—	—
318	(Brewers rice)		100.0	—	—	—	—	—
319	groats, polished (Rice, polished)	4-03-942	89.0	—	—	—	4.0	—
320			100.0	—	—	—	4.0	—
321	polishings	4-03-943	90.0	—	—	—	90.0	—
322			100.0	—	—	—	100.0	—
	RYE *Secale cereale*							
323	distillers grains, dehy	5-04-023	92.0	—	—	—	—	—
324			100.0	—	—	—	—	—
325	flour by-product, less than 8.5%	4-04-031	89.0	—	—	—	—	—
326	fiber (Rye middlings)		100.0	—	—	—	—	—
327	fresh	2-04-018	24.0	82.0	—	—	—	—
328			100.0	343.0	—	—	—	—
329	grain	4-04-047	88.0	0.0	—	—	15.0	—
330			100.0	0.0	—	—	17.0	—
331	silage	3-04-020	32.0	19.0	—	—	—	—
332			100.0	58.0	—	—	—	—
	RYEGRASS, PERENNIAL *Lolium perenne*							
333	fresh	2-04-073	25.0	98.0	—	—	—	—
334			100.0	401.0	—	—	—	—
335	hay, sun-cured, late vegetative	1-04-065	86.0	248.0	—	—	—	—
336			100.0	290.0	—	—	—	—
	RYEGRASS, PERENNIAL *Lolium perenne*							
337	fresh	2-04-086	27.0	59.0	—	—	47.0	—
338			100.0	222.0	—	—	178.0	—
339	hay, sun-cured	1-04-077	86.0	103.0	—	—	182.0	—
340			100.0	120.0	—	—	211.0	—

Entry Num- ber	Water Soluble Vitamins Bio- tin (mg/kg)	Cho- line (mg/kg)	Folic Acid (Folacin) (mg/kg)	Niacin (mg/kg)	Pantothe- nic Acid (mg/kg)	Ribo- flavin (mg/kg)	Thia- mine (mg/kg)	Vita- min B$_6$ (mg/kg)	Vita- min B$_{12}$ (μg/kg)	Xan- tho- phylls (mg/kg)
303	—	—	—	—	—	—	—	—	—	—
304	—	—	—	—	—	—	—	—	—	—
305	—	—	—	—	—	—	—	—	—	—
306	—	—	—	—	—	—	—	—	—	—
307	—	6,533.0	—	155.0	9.0	3.0	1.8	—	—	—
308	—	7,103.0	—	168.0	9.8	3.3	1.9	—	—	—
309	—	6,625.0	—	147.0	8.0	5.8	1.6	7.2	—	—
310	—	7,278.0	—	161.0	8.8	6.4	1.7	8.0	—	—
311	—	—	—	—	—	—	—	—	—	—
312	—	—	—	—	—	—	—	—	—	—
313	0.43	1,230.0	2.2	300.0	22.8	2.6	22.4	—	—	—
314	0.47	1,357.0	2.4	330.0	25.2	2.8	24.7	—	—	—
315	0.08	957.0	0.4	34.0	8.1	1.0	2.9	4.4	—	—
316	0.09	1,076.0	0.4	39.0	9.1	1.2	3.2	5.0	—	—
317	—	877.0	—	23.0	3.3	0.4	1.4	—	—	—
318	—	991.0	—	26.0	3.7	0.5	1.6	—	—	—
319	—	902.0	0.1	15.0	3.5	0.6	0.7	0.4	—	—
320	—	1,018.0	0.2	17.0	3.9	0.6	0.7	0.4	—	—
321	0.62	1,249.0	—	506.0	46.4	1.8	20.0	—	—	—
322	0.68	1,383.0	—	560.0	51.4	2.0	22.1	—	—	—
323	—	—	—	17.0	5.2	3.3	1.3	—	—	—
324	—	—	—	18.0	5.7	3.6	1.4	—	—	—
325	—	—	—	17.0	22.9	2.4	3.3	—	—	—
326	—	—	—	19.0	25.7	2.7	3.7	—	—	—
327	—	—	—	—	—	—	—	—	—	—
328	—	—	—	—	—	—	—	—	—	—
329	0.06	419.0	0.6	19.0	8.0	1.6	3.6	2.6	—	—
330	0.06	479.0	0.7	21.0	9.1	1.9	4.2	2.9	—	—
331	—	—	—	—	—	—	—	—	—	—
332	—	—	—	—	—	—	—	—	—	—
333	—	—	—	—	—	—	—	—	—	—
334	—	—	—	—	—	—	—	—	—	—
335	—	—	—	—	—	—	—	—	—	—
336	—	—	—	—	—	—	—	—	—	—
337	—	—	—	—	—	—	—	—	—	—
338	—	—	—	—	—	—	—	—	—	—
339	—	—	—	—	—	—	—	—	—	—
340	—	—	—	—	—	—	—	—	—	—

TABLE 3 Composition of Important Feeds: Vitamins—*Continued*

Entry Number	Feed Name Description	International Feed Number	Dry Matter (%)	Carotene (Provitamin A) (mg/kg)	Vitamin A (IU/g)	Vitamin D₂ (IU/g)	Vitamin E (mg/kg)	Vitamin K (mg/kg)
				Fat-Soluble Vitamins				
	SAFFLOWER *Carthamus tinctorius*							
341	seeds, meal mech extd	5-04-109	91.0	—	—	—	1.0	—
342			100.0	—	—	—	1.0	—
343	seeds, meal solv extd	5-04-110	92.0	—	—	—	1.0	—
344			100.0	—	—	—	1.0	—
345	seeds without hulls, meal solv	5-07-959	92.0	—	—	—	1.0	—
346	extd		100.0	—	—	—	1.0	—
	SAGEBRUSH, BIG *Artemisia tridentata*							
347	browse, fresh, stem-cured	2-07-992	65.0	10.0	—	—	—	—
348			100.0	16.0	—	—	—	—
	SAGEBRUSH, BUD *Artemisia spinescens*							
349	browse, fresh, early vegetative	2-07-991	23.0	5.0	—	—	—	—
350			100.0	24.0	—	—	—	—
	SALTBUSH, NUTTALL *Atriplex nuttallii*							
351	browse, fresh, stem-cured	2-07-993	55.0	10.0	—	—	—	—
352			100.0	19.0	—	—	—	—
	SCREENINGS—SEE BARLEY, SEE CEREALS, SEE WHEAT							
	SEAWEED, KELP *Laminariales* (order)–*Fucales* (order)							
353	whole, dehy	4-08-073	91.0	—	—	—	—	—
354			100.0	—	—	—	—	—
	SESAME *Sesamum indicum*							
355	seeds, meal mech extd	5-04-220	93.0	0.0	—	—	—	—
356			100.0	0.0	—	—	—	—
	SORGHUM *Sorghum bicolor*							
357	aerial part with heads, sun-cured	1-07-960	89.0	46.0	—	—	—	—
358	(Fodder)		100.0	52.0	—	—	—	—
359	distillers grains, dehy	5-04-374	94.0	—	—	—	—	—
360			100.0	—	—	—	—	—
361	grain	4-04-383	90.0	1.0	—	26.0	10.0	0.2
362			100.0	1.0	—	29.0	12.0	0.2
363	grain, 8–10% protein	4-20-893	87.0	—	—	—	7.0	—
364			100.0	—	—	—	8.0	—
365	silage	3-04-323	30.0	5.0	—	196.0	—	—
366			100.0	15.0	—	662.0	—	—
	SORGHUM, JOHNSONGRASS *Sorghum halepense*							
367	hay, sun-cured	1-04-407	89.0	35.0	—	—	—	—
368			100.0	39.0	—	—	—	—
	SORGHUM, SORGO *Sorghum bicolor saccharatum*							
369	silage	3-04-468	27.0	10.0	—	—	—	—
370			100.0	36.0	—	—	—	—

| Entry Number | Water Soluble Vitamins | | | | | | | | | |
	Bio-tin (mg/kg)	Cho-line (mg/kg)	Folic Acid (Folacin) (mg/kg)	Niacin (mg/kg)	Pantothe-nic Acid (mg/kg)	Ribo-flavin (mg/kg)	Thia-mine (mg/kg)	Vita-min B$_6$ (mg/kg)	Vita-min B$_{12}$ (μg/kg)	Xan-tho-phylls (mg/kg)
341	1.41	1,178.0	0.4	—	—	—	—	—	—	—
342	1.54	1,287.0	0.5	—	—	—	—	—	—	—
343	1.43	820.0	0.5	11.0	33.9	2.3	—	—	—	—
344	1.56	889.0	0.5	12.0	36.7	2.5	—	—	—	—
345	1.67	3,248.0	1.6	22.0	39.1	2.4	4.5	11.3	—	—
346	1.82	3,543.0	1.7	24.0	42.7	2.6	4.9	12.4	—	—
347	—	—	—	—	—	—	—	—	—	—
348	—	—	—	—	—	—	—	—	—	—
349	—	—	—	—	—	—	—	—	—	—
350	—	—	—	—	—	—	—	—	—	—
351	—	—	—	—	—	—	—	—	—	—
352	—	—	—	—	—	—	—	—	—	—
353	—	—	—	—	—	—	—	—	—	350.0
354	—	—	—	—	—	—	—	—	—	383.0
355	—	1,535.0	—	19.0	6.0	3.4	2.8	12.5	—	—
356	—	1,655.0	—	20.0	6.4	3.6	3.0	13.4	—	—
357	—	—	—	—	—	—	—	—	—	—
358	—	—	—	—	—	—	—	—	—	—
359	0.34	788.0	—	54.0	5.7	2.9	0.7	—	—	—
360	0.37	841.0	—	58.0	6.1	3.1	0.7	—	—	—
361	0.38	661.0	0.2	39.0	11.2	1.2	4.2	4.5	—	—
362	0.42	737.0	0.2	43.0	12.5	1.4	4.7	5.0	—	—
363	0.26	668.0	0.2	41.0	12.4	1.3	3.9	5.2	—	—
364	0.30	769.0	0.3	48.0	14.3	1.5	4.5	6.0	—	—
365	—	—	—	—	—	—	—	—	—	—
366	—	—	—	—	—	—	—	—	—	—
367	—	—	—	—	—	—	—	—	—	—
368	—	—	—	—	—	—	—	—	—	—
369	—	—	—	—	—	—	—	—	—	—
370	—	—	—	—	—	—	—	—	—	—

TABLE 3   Composition of Important Feeds: Vitamins—*Continued*

Entry Number	Feed Name Description	International Feed Number	Dry Matter (%)	Carotene (Provitamin A) (mg/kg)	Vitamin A (IU/g)	Vitamin D₂ (IU/g)	Vitamin E (mg/kg)	Vitamin K (mg/kg)
				Fat-Soluble Vitamins				
	SORGHUM, SUDANGRASS							
	Sorghum bicolor sudanense							
371	fresh, early vegetative	2-04-484	18.0	35.0	—	—	—	—
372			100.0	198.0	—	—	—	—
373	fresh, midbloom	2-04-485	23.0	42.0	—	—	—	—
374			100.0	183.0	—	—	—	—
375	hay, sun-cured	1-04-480	91.0	54.0	—	—	—	—
376			100.0	59.0	—	—	—	—
377	silage	3-04-499	28.0	30.0	—	—	—	—
378			100.0	105.0	—	—	—	—
	SOYBEAN *Glycine max*							
379	flour by-product (Soybean mill	4-04-594	90.0	—	—	—	—	—
380	feed)		100.0	—	—	—	—	—
381	hay, sun-cured	1-04-558	89.0	41.0	—	947.0	26.0	—
382			100.0	45.0	—	1,059.0	30.0	—
383	protein concentrate, more than	5-08-038	92.0	—	—	—	—	0.0
384	70% protein		100.0	—	—	—	—	0.0
385	seeds	5-04-610	92.0	1.0	—	—	33.0	—
386			100.0	1.0	—	—	37.0	—
387	seeds, heat processed	5-04-597	90.0	—	—	—	—	—
388			100.0	—	—	—	—	—
389	seeds, meal mech extd	5-04-600	90.0	0.0	—	—	7.0	—
390			100.0	0.0	—	—	7.0	—
391	seeds, meal solv extd	5-04-604	90.0	0.0	—	—	3.0	—
392			100.0	0.0	—	—	3.0	—
393	seeds without hulls, meal solv	5-04-612	90.0	—	—	—	2.0	—
394	extd		100.0	—	—	—	3.0	—
	SUDANGRASS—SEE SORGHUM, SUDANGRASS							
	SUGARCANE *Saccharum officinarum*							
	molasses—see Molasses and syrup							
	SUNFLOWER, COMMON							
	Helianthus annuus							
395	seeds, meal solv extd	5-09-340	90.0	—	—	—	—	—
396			100.0	—	—	—	—	—
397	seeds without hulls, meal solv	5-04-739	93.0	—	—	—	11.0	—
398	extd		100.0	—	—	—	12.0	—
	SWEETCLOVER, YELLOW							
	Melilotus officinalis							
399	hay, sun-cured	1-04-754	87.0	86.0	—	1,636.0	—	—
400			100.0	99.0	—	1,874.0	—	—
	SWINE *Sus scrofa*							
401	livers, fresh	5-04-792	30.0	—	109.4	—	—	—
402			100.0	—	361.7	—	—	—
403	lungs, fresh	5-26-140	16.0	—	0.8	—	4.0	—
404			100.0	—	4.8	—	27.0	—
	TIMOTHY *Phleum pratense*							
405	fresh	2-04-912	30.0	54.0	—	—	34.0	—
406			100.0	179.0	—	—	111.0	—

Entry Num-ber	Water Soluble Vitamins									
	Bio-tin (mg/kg)	Cho-line (mg/kg)	Folic Acid (Folacin) (mg/kg)	Niacin (mg/kg)	Pantothe-nic Acid (mg/kg)	Ribo-flavin (mg/kg)	Thia-mine (mg/kg)	Vita-min B_6 (mg/kg)	Vita-min B_{12} (μg/kg)	Xan-tho-phylls (mg/kg)
371	—	—	—	—	—	—	—	—	—	—
372	—	—	—	—	—	—	—	—	—	—
373	—	—	—	—	—	—	—	—	—	—
374	—	—	—	—	--	—	—	—	—	—
375	—	—	—	—	—	—	—	—	—	—
376	—	—	—	—	—	—	—	—	—	—
377	—	—	—	—	—	—	—	—	—	—
378	—	—	—	—	—	—	—	—	—	—
379	0.22	492.0	0.2	24.0	13.2	3.5	2.2	2.2	—	—
380	0.25	549.0	0.3	27.0	14.7	3.9	2.5	2.5	—	—
381	—	—	—	—	—	—	—	—	—	—
382	—	—	—	—	—	—	—	—	—	—
383	—	2.0	—	5.0	3.5	0.7	0.3	—	—	—
384	—	2.0	—	5.0	3.8	0.8	0.4	—	—	—
385	0.34	2,690.0	3.6	22.0	15.8	2.8	9.7	—	—	—
386	0.37	2,939.0	3.9	24.0	17.3	3.1	10.6	—	—	—
387	0.29	—	3.5	22.0	15.6	2.6	—	—	—	—
388	0.32	—	3.9	24.0	17.4	2.9	—	—	—	—
389	0.33	2,623.0	6.4	31.0	14.3	3.4	3.9	—	—	—
390	0.36	2,916.0	7.1	34.0	15.8	3.8	4.3	—	—	—
391	0.32	2,614.0	0.7	28.0	16.3	2.9	5.6	6.0	—	0.0
392	0.36	2,915.0	0.7	31.0	18.2	3.2	6.2	6.7	—	0.0
393	0.32	2,753.0	0.7	22.0	14.8	2.9	3.1	4.9	—	0.0
394	0.36	3,054.0	0.8	24.0	16.4	3.2	3.4	5.5	—	0.0
395	—	3,791.0	—	264.0	29.9	3.0	3.0	11.1	—	—
396	—	4,214.0	—	293.0	33.3	3.4	3.4	12.4	—	—
397	—	4,120.0	—	268.0	40.8	3.9	3.1	13.7	—	—
398	—	4,430.0	—	288.0	43.9	4.2	3.4	14.8	—	—
399	—	—	—	—	—	—	—	—	—	—
400	—	—	—	—	—	—	—	—	—	—
401	0.75	—	2.1	165.0	23.6	27.3	2.3	3.0	283.0	—
402	2.49	—	6.9	544.0	77.9	90.3	7.7	10.0	935.0	—
403	0.05	2,271.0	0.1	13.0	0.6	2.1	0.4	0.4	24.0	—
404	0.32	14,373.0	0.9	80.0	4.1	13.4	2.2	2.2	152.0	—
405	—	—	—	—	—	3.5	0.9	—	—	—
406	—	—	—	—	—	11.5	2.9	—	—	—

TABLE 3   Composition of Important Feeds: Vitamins—*Continued*

Entry Number	Feed Name Description	International Feed Number	Dry Matter (%)	Carotene (Provitamin A) (mg/kg)	Vitamin A (IU/g)	Vitamin D₂ (IU/g)	Vitamin E (mg/kg)	Vitamin K (mg/kg)
407	hay, sun-cured	1-04-893	90.0	25.0	—	1,930.0	34.0	—
408			100.0	28.0	—	2,138.0	38.0	—
409	silage	3-04-922	34.0	31.0	—	—	—	—
410			100.0	90.0	—	—	—	—
	TOMATO *Lycopersicon esculentum*							
411	pomace, dehy	5-05-041	92.0	—	—	—	—	—
412			100.0	—	—	—	—	—
	TORULA DRIED YEAST—SEE YEAST, TORULA							
	TREFOIL, BIRDSFOOT *Lotus corniculatus*							
413	hay, sun-cured	1-05-044	92.0	173.0	—	1,421.0	—	—
414			100.0	188.0	—	1,544.0	—	—
	TRITICALE *Triticale hexaploide*							
415	grain	4-20-362	90.0	—	—	—	—	—
416			100.0	—	—	—	—	—
	TURNIP *Brassica rapa rapa*							
417	roots, fresh	4-05-067	9.0	—	—	—	—	—
418			100.0	—	—	—	—	—
	VETCH *Vicia* spp							
419	hay, sun-cured	1-05-106	89.0	411.0	—	—	—	—
420			100.0	461.0	—	—	—	—
	WHEAT *Triticum aestivum*							
421	bran	4-05-190	89.0	3.0	—	—	18.0	—
422			100.0	3.0	—	—	21.0	—
423	flour, hard red spring, less than	4-08-113	88.0	—	—	—	2.0	—
424	1.5% fiber		100.0	—	—	—	3.0	—
425	flour, less than 1.5% fiber	4-05-199	88.0	—	—	—	2.0	—
426	(Wheat feed flour)		100.0	—	—	—	3.0	—
427	flour by-product, less than 4%	4-05-203	88.0	—	—	—	33.0	—
428	fiber (Wheat red dog)		100.0	—	—	—	37.0	—
429	flour by-product, less than 7%	4-05-201	88.0	—	—	—	54.0	—
430	fiber (Wheat shorts)		100.0	—	—	—	61.0	—
431	fresh	2-08-078	25.0	78.0	—	—	—	—
432			100.0	316.0	—	—	—	—
433	germs, ground	5-05-218	88.0	—	—	—	142.0	—
434			100.0	—	—	—	160.0	—
435	grain	4-05-211	89.0	0.0	—	—	15.0	—
436			100.0	0.0	—	—	17.0	—
437	grain, hard red spring	4-05-258	88.0	0.0	—	—	13.0	—
438			100.0	0.0	—	—	14.0	—
439	grain, hard red winter	4-05-268	88.0	—	—	—	11.0	—
440			100.0	—	—	—	12.0	—
441	grain, soft red winter	4-05-294	88.0	—	—	—	16.0	—
442			100.0	—	—	—	18.0	—
443	grain, soft white winter	4-05-337	89.0	—	—	—	18.0	—
444			100.0	—	—	—	20.0	—
445	hay, sun-cured	1-05-172	88.0	75.0	—	1,352.0	—	—
446			100.0	85.0	—	1,544.0	—	—

| Entry Num- ber | Water Soluble Vitamins | | | | | | | | | |
	Bio- tin (mg/kg)	Cho- line (mg/kg)	Folic Acid (Folacin) (mg/kg)	Niacin (mg/kg)	Pantothe- nic Acid (mg/kg)	Ribo- flavin (mg/kg)	Thia- mine (mg/kg)	Vita- min B_6 (mg/kg)	Vita- min B_{12} (μg/kg)	Xan- tho- phylls (mg/kg)
407	0.06	732.0	2.1	26.0	7.1	9.1	1.5	—	—	—
408	0.07	811.0	2.3	29.0	7.9	10.1	1.7	—	—	—
409	—	—	—	—	—	—	—	—	—	—
410	—	—	—	—	—	—	—	—	—	—
411	—	—	—	—	—	6.1	11.3	—	—	—
412	—	—	—	—	—	6.7	12.3	—	—	—
413	—	—	—	—	—	14.8	6.3	—	—	—
414	—	—	—	—	—	16.1	6.8	—	—	—
415	—	462.0	—	—	—	0.4	—	—	—	—
416	—	514.0	—	—	—	0.5	—	—	—	—
417	—	—	0.3	7.0	1.8	0.6	0.7	—	—	—
418	—	—	2.8	72.0	19.0	6.5	7.1	—	—	—
419	—	—	—	—	—	—	—	—	—	—
420	—	—	—	—	—	—	—	—	—	—
421	0.29	1,596.0	1.4	238.0	29.7	4.1	7.0	8.5	—	—
422	0.32	1,797.0	1.6	268.0	33.5	4.6	7.9	9.6	—	—
423	—	952.0	0.1	13.0	6.8	0.6	1.9	0.8	—	—
424	—	1,076.0	0.1	14.0	7.7	0.6	2.1	0.9	—	—
425	—	829.0	0.1	12.0	6.1	0.5	1.8	0.9	—	—
426	—	947.0	0.1	14.0	7.0	0.6	2.1	1.0	—	—
427	0.11	1,534.0	0.8	42.0	13.3	2.2	22.8	4.6	—	—
428	0.12	1,742.0	0.9	48.0	15.1	2.5	25.9	5.2	—	—
429	—	1,813.0	1.7	107.0	22.3	4.2	19.1	7.2	—	—
430	—	2,050.0	1.9	121.0	25.3	4.7	21.7	8.2	—	—
431	—	—	—	11.0	5.2	4.9	0.9	—	—	—
432	—	—	—	44.0	21.2	19.8	3.5	—	—	—
433	0.22	3,062.0	2.2	72.0	20.1	6.0	22.8	11.4	—	—
434	0.24	3,468.0	2.4	81.0	22.8	6.8	25.8	12.9	—	—
435	0.10	964.0	0.4	57.0	10.2	1.4	4.3	4.9	1.0	—
436	0.11	1,085.0	0.5	64.0	11.4	1.6	4.8	5.6	1.0	—
437	0.11	1,051.0	0.4	57.0	9.8	1.4	4.2	5.1	—	—
438	0.13	1,200.0	0.5	65.0	11.2	1.6	4.8	5.8	—	—
439	0.11	1,041.0	0.4	54.0	9.8	1.4	4.2	3.0	—	—
440	0.12	1,179.0	0.4	61.0	11.1	1.6	4.8	3.4	—	—
441	—	929.0	0.4	52.0	9.6	1.5	4.5	3.2	—	—
442	—	1,053.0	0.5	59.0	10.9	1.7	5.1	3.6	—	—
443	0.11	978.0	0.4	53.0	11.2	1.2	4.7	4.1	—	—
444	0.12	1,097.0	0.4	59.0	12.6	1.3	5.3	4.6	—	—
445	—	—	—	—	—	14.9	—	—	—	—
446	—	—	—	—	—	17.0	—	—	—	—

TABLE 3 Composition of Important Feeds: Vitamins—*Continued*

Entry Number	Feed Name Description	International Feed Number	Dry Matter (%)	Fat-Soluble Vitamins				
				Carotene (Provita-min A) (mg/kg)	Vita-min A (IU/g)	Vita-min D$_2$ (IU/g)	Vita-min E (mg/kg)	Vita-min K (mg/kg)
447	mill run, less than 9.5% fiber	4-05-206	90.0	—	—	—	—	—
448			100.0	—	—	—	—	—
449	silage, early vegetative	3-05-184	30.0	44.0	—	—	—	—
450			100.0	147.0	—	—	—	—
451	straw	1-05-175	89.0	2.0	—	587.0	—	—
452			100.0	2.0	—	662.0	—	—
	WHEAT, DURUM *Triticum durum*							
453	grain	4-05-224	88.0	—	—	—	—	—
454			100.0	—	—	—	—	—
	WHEATGRASS, CRESTED *Agropyron desertorum*							
455	fresh	2-05-429	39.0	83.0	—	—	—	—
456			100.0	213.0	—	—	—	—
	WHEY							
457	dehy (Cattle)	4-01-182	93.0	—	0.5	—	0.0	—
458			100.0	—	0.5	—	0.0	—
459	low lactose, dehy (Dried whey	4-01-186	93.0	—	—	—	—	—
460	product) (Cattle)		100.0	—	—	—	—	—
	YEAST *Saccharomyces cerevisiae*							
461	brewers, dehy	7-05-527	93.0	—	—	—	2.0	—
462			100.0	—	—	—	2.0	—
	YEAST, TORULA *Torulopsis utilis*							
463	torula, dehy	7-05-534	93.0	—	—	—	—	—
464			100.0	—	—	—	—	—

Entry Num-ber	Water Soluble Vitamins Bio-tin (mg/kg)	Cho-line (mg/kg)	Folic Acid (Folacin) (mg/kg)	Niacin (mg/kg)	Pantothe-nic Acid (mg/kg)	Ribo-flavin (mg/kg)	Thia-mine (mg/kg)	Vita-min B₆ (mg/kg)	Vita-min B₁₂ (μg/kg)	Xan-tho-phylls (mg/kg)
447	0.31	1,005.0	1.1	116.0	13.7	2.1	15.3	11.0	—	—
448	0.34	1,118.0	1.2	129.0	15.2	2.4	17.0	12.2	—	—
449	—	—	—	—	—	—	—	—	—	—
450	—	—	—	—	—	—	—	—	—	—
451	—	—	—	—	—	2.2	—	—	—	—
452	—	—	—	—	—	2.4	—	—	—	—
453	—	—	0.4	52.0	8.9	1.1	6.4	3.0	—	—
454	—	—	0.4	60.0	10.1	1.2	7.3	3.4	—	—
455	—	842.0	—	—	—	3.4	1.3	—	—	—
456	—	2,161.0	—	—	—	8.6	3.3	—	—	—
457	0.35	1,793.0	0.9	11.0	46.3	27.5	4.0	3.3	19.0	—
458	0.38	1,921.0	0.9	11.0	49.6	29.4	4.3	3.6	20.0	—
459	0.50	3,859.0	0.7	18.0	75.0	48.6	5.0	4.9	35.0	—
460	0.54	4,133.0	0.8	19.0	80.3	52.1	5.4	5.3	38.0	—
461	1.01	3,949.0	9.6	450.0	110.7	35.6	92.7	37.1	1.0	—
462	1.08	4,227.0	10.3	482.0	118.4	38.1	99.2	39.8	1.0	—
463	1.37	3,005.0	24.2	489.0	93.8	44.4	6.2	36.3	4.0	—
464	1.47	3,223.0	26.0	525.0	100.6	47.6	6.6	38.9	4.0	—

TABLE 4   Composition of Important Feeds: Amino Acids, Data Expressed As-Fed and Dry (100% Dry Matter)

Entry Number	Feed Name Description	International Feed Number	Dry Matter (%)	Crude Protein (%)	Arginine (%)	Glycine (%)	Histidine (%)	Isoleucine (%)
	ALFALFA *Medicago sativa*							
001	hay, sun-cured	1-00-078	90.0	16.4	0.72	0.70	0.30	0.73
002			100.0	18.2	0.81	0.78	0.33	0.81
003	hay, sun-cured, early vegetative	1-00-050	90.0	20.7	1.14	1.03	0.50	0.96
004			100.0	23.0	1.27	1.14	0.55	1.07
005	hay, sun-cured, late vegetative	1-00-054	90.0	17.9	0.84	0.84	0.38	0.79
006			100.0	20.0	0.94	0.94	0.42	0.88
007	hay, sun-cured, early bloom	1-00-059	90.0	16.2	0.73	0.68	0.34	0.60
008			100.0	18.0	0.81	0.75	0.38	0.67
009	hay, sun-cured, full bloom	1-00-068	90.0	13.5	0.67	0.69	0.32	0.61
010			100.0	15.0	0.74	0.77	0.35	0.68
011	leaves, sun-cured	1-00-146	89.0	20.6	1.16	—	0.36	0.89
012			100.0	23.1	1.30	—	0.40	1.00
013	meal dehy, 15% protein	1-00-022	90.0	15.6	0.59	0.70	0.27	0.64
014			100.0	17.3	0.65	0.78	0.30	0.71
015	meal dehy, 17% protein	1-00-023	92.0	17.3	0.77	0.84	0.33	0.81
016			100.0	18.9	0.84	0.91	0.36	0.88
017	meal dehy, 20% protein	1-00-024	92.0	20.2	0.96	0.98	0.37	0.89
018			100.0	22.0	1.05	1.07	0.41	0.97
019	meal dehy, 22% protein	1-07-851	93.0	22.2	0.96	1.09	0.44	1.06
020			100.0	23.9	1.04	1.18	0.47	1.15
	BAKERY							
021	waste, dehy (Dried bakery	4-00-466	92.0	9.8	0.47	0.82	0.13	0.45
022	product)		100.0	10.7	0.51	0.89	0.14	0.49
	BARLEY *Hordeum vulgare*							
023	grain	4-00-549	88.0	11.9	0.51	0.38	0.24	0.45
024			100.0	13.5	0.58	0.43	0.28	0.51
025	grain, Pacific Coast	4-07-939	89.0	9.6	0.44	0.30	0.21	0.40
026			100.0	10.8	0.50	0.34	0.23	0.45
027	grain screenings	4-00-542	89.0	11.7	—	—	—	—
028			100.0	13.1	—	—	—	—
029	malt sprouts, dehy	5-00-545	94.0	26.3	1.12	1.12	0.52	1.11
030			100.0	28.1	1.19	1.20	0.56	1.19
	BEAN, NAVY *Phaseolus vulgaris*							
031	seeds	5-00-623	89.0	22.6	1.19	0.80	—	—
032			100.0	25.3	1.33	0.89	—	—
	BEET, SUGAR *Beta vulgaris altissima*							
033	pulp, dehy	4-00-669	91.0	8.8	0.30	—	0.20	0.30
034			100.0	9.7	0.33	—	0.22	0.33
035	pulp with molasses, dehy	4-00-672	92.0	9.3	0.30	—	—	—
036			100.0	10.1	0.33	—	—	—
	BLOOD							
037	meal	5-00-380	92.0	79.8	3.25	3.42	3.97	0.87
038			100.0	87.2	3.55	3.74	4.34	0.95
039	meal flash dehy	5-26-006	92.0	85.9	3.34	4.23	4.57	0.88
040			100.0	93.3	3.63	4.60	4.97	0.96
041	meal spray dehy (Blood flour)	5-00-381	93.0	86.5	3.60	3.85	5.20	0.91
042			100.0	93.0	3.88	4.14	5.59	0.98
	BREWERS							
043	grains, dehy	5-02-141	92.0	27.1	1.27	1.08	0.52	1.54
044			100.0	29.4	1.38	1.18	0.56	1.68

Entry Number	Leucine (%)	Lysine (%)	Methio- nine (%)	Cystine (%)	Phenyl- alanine (%)	Tyrosine (%)	Serine (%)	Threo- nine (%)	Trypto- phan (%)	Valine (%)
001	1.12	0.75	0.18	0.24	0.69	0.46	0.69	0.63	0.21	0.74
002	1.25	0.84	0.20	0.27	0.76	0.51	0.76	0.70	0.24	0.82
003	1.64	1.27	0.36	—	1.07	0.74	0.97	1.08	—	1.22
004	1.82	1.41	0.40	—	1.19	0.82	1.08	1.20	—	1.35
005	1.37	0.99	0.27	—	0.83	0.56	0.77	—	—	0.97
006	1.53	1.10	0.30	—	0.93	0.62	0.86	—	—	1.08
007	1.07	0.81	0.19	0.31	0.71	0.48	0.65	0.60	—	0.79
008	1.19	0.90	0.21	0.34	0.78	0.53	0.72	0.66	—	0.88
009	1.05	0.78	0.20	—	0.68	0.46	0.64	0.55	—	0.77
010	1.17	0.87	0.22	—	0.75	0.51	0.71	0.61	—	0.86
011	1.34	0.98	0.36	0.36	0.89	—	—	0.72	0.45	0.98
012	1.50	1.10	0.40	0.40	1.00	—	—	0.80	0.50	1.10
013	1.02	0.59	0.22	0.21	0.62	0.41	0.60	0.56	0.38	0.75
014	1.13	0.66	0.24	0.23	0.69	0.45	0.67	0.62	0.42	0.83
015	1.28	0.85	0.27	0.29	0.80	0.54	0.71	0.71	0.34	0.88
016	1.39	0.93	0.29	0.31	0.87	0.59	0.77	0.77	0.37	0.96
017	1.41	0.90	0.32	0.32	0.94	0.62	0.86	0.81	0.41	1.04
018	1.54	0.98	0.34	0.35	1.03	0.67	0.94	0.88	0.45	1.13
019	1.63	0.97	0.34	0.30	1.13	0.64	0.97	0.97	0.49	1.29
020	1.75	1.05	0.37	0.32	1.22	0.69	1.05	1.04	0.52	1.39
021	0.73	0.31	0.17	0.17	0.40	0.41	—	0.49	0.10	0.42
022	0.80	0.34	0.19	0.18	0.44	0.45	—	0.53	0.11	0.46
023	0.75	0.39	0.15	0.21	0.58	0.34	0.43	0.37	0.15	0.57
024	0.85	0.44	0.17	0.24	0.66	0.38	0.49	0.42	0.17	0.64
025	0.60	0.26	0.14	0.20	0.48	0.31	0.32	0.31	0.12	0.46
026	0.67	0.30	0.16	0.22	0.53	0.34	0.36	0.35	0.14	0.52
027	—	—	—	—	—	—	—	0.36	—	—
028	—	—	—	—	—	—	—	0.40	—	—
029	1.65	1.21	0.33	0.24	0.92	0.61	—	1.01	0.42	1.45
030	1.76	1.29	0.35	0.25	0.98	0.65	—	1.07	0.44	1.55
031	—	1.29	0.25	0.23	—	—	—	—	0.24	—
032	—	1.44	0.28	0.26	—	—	—	—	0.27	—
033	0.60	0.60	0.01	0.01	0.30	0.40	—	0.40	0.10	0.40
034	0.66	0.66	0.01	0.01	0.33	0.44	—	0.44	0.11	0.44
035	—	0.60	0.01	—	—	—	—	0.25	0.09	—
036	—	0.65	0.01	—	—	—	—	0.27	0.10	—
037	9.94	6.33	0.88	1.24	5.49	1.92	4.35	3.56	0.98	6.52
038	10.86	6.92	0.97	1.35	6.00	2.09	4.75	3.89	1.07	7.12
039	11.48	7.56	0.95	1.20	6.41	2.32	5.46	4.07	1.06	8.03
040	12.47	8.21	1.03	1.30	6.97	2.52	5.93	4.42	1.15	8.72
041	11.03	7.48	0.88	0.72	5.92	2.27	3.55	3.65	1.05	7.56
042	11.86	8.04	0.95	0.78	6.36	2.44	3.82	3.93	1.13	8.13
043	2.49	0.88	0.46	0.35	1.44	1.20	1.30	0.93	0.37	1.61
044	2.70	0.95	0.50	0.38	1.56	1.30	1.42	1.01	0.40	1.75

TABLE 4 Composition of Important Feeds: Amino Acids—*Continued*

Entry Number	Feed Name Description	International Feed Number	Dry Matter (%)	Crude Protein (%)	Arginine (%)	Glycine (%)	Histidine (%)	Isoleucine (%)
	BROOMCORN MILLET—SEE MILLET, PROSO							
	BUCKWHEAT, COMMON *Fagopyrum sagittatum*							
045	grain	4-00-994	88.0	11.0	0.98	0.71	0.26	0.36
046			100.0	12.5	1.12	0.81	0.30	0.41
	BUTTERMILK							
047	dehy (Cattle)	5-01-160	92.0	31.7	1.08	0.47	0.85	2.42
048			100.0	34.4	1.17	0.51	0.92	2.62
	CARROT *Daucus* spp							
049	roots, fresh	4-01-145	12.0	1.2	0.15	—	0.03	0.17
050			100.0	9.9	1.27	—	0.25	1.44
	CASEIN							
051	dehy (Cattle)	5-01-162	91.0	84.0	3.49	1.61	2.59	5.72
052			100.0	92.7	3.85	1.77	2.86	6.32
053	precipitated dehy	5-20-837	92.0	85.0	3.42	1.81	2.52	4.77
054			100.0	92.4	3.72	1.97	2.74	5.19
	CASSAVA, COMMON *Manihot esculenta*							
055	tubers, dehy	4-09-598	88.0	2.3	0.29	0.01	0.07	0.03
056			100.0	2.6	0.33	0.01	0.08	0.03
	CATTLE *Bos taurus* buttermilk—see Buttermilk							
057	lungs, fresh	5-07-941	21.0	13.9	0.66	1.19	0.24	0.29
058			100.0	65.0	3.11	5.57	1.13	1.37
059	manure, without bedding, dehy	1-01-190	94.0	15.2	0.28	0.49	0.17	0.36
060			100.0	16.1	0.29	0.51	0.18	0.38
	milk—see Milk skim milk—see Milk							
	CEREALS							
061	screenings	4-02-156	90.0	12.1	0.95	0.40	0.30	0.50
062			100.0	13.4	1.06	0.44	0.34	0.56
063	screenings refuse	4-02-151	91.0	12.8	0.68	0.59	0.30	0.52
064			100.0	14.1	0.75	0.65	0.33	0.58
065	screenings uncleaned	4-02-153	92.0	13.9	0.67	0.61	0.30	0.45
066			100.0	15.1	0.73	0.66	0.33	0.49
	CHICKEN *Gallus domesticus*							
067	hens, whole, fresh	5-07-950	33.0	19.9	0.85	1.10	0.25	0.66
068			100.0	60.3	2.59	3.32	0.77	2.00
	CITRUS *Citrus* spp							
069	pulp fines (Dried citrus meal)	4-01-235	91.0	6.5	0.28	—	—	—
070			100.0	7.1	0.31	—	—	—
071	pulp without fines, dehy (Dried citrus pulp)	4-01-237	91.0	6.1	0.24	—	—	—
072			100.0	6.7	0.27	—	—	—
	syrup—see Molasses and syrup							
	CLOVER, LADINO *Trifolium repens*							
073	hay, sun-cured	1-01-378	90.0	19.7	0.99	0.90	0.45	1.08
074			100.0	22.0	1.10	1.00	0.50	1.20

Entry Number	Leucine (%)	Lysine (%)	Methionine (%)	Cystine (%)	Phenylalanine (%)	Tyrosine (%)	Serine (%)	Threonine (%)	Tryptophan (%)	Valine (%)
045	0.55	0.61	0.19	0.20	0.44	—	—	0.45	0.18	0.53
046	0.63	0.70	0.22	0.23	0.50	—	—	0.51	0.21	0.61
047	3.21	2.28	0.71	0.39	1.46	1.00	1.50	1.52	0.49	2.58
048	3.48	2.47	0.76	0.42	1.58	1.08	1.62	1.64	0.53	2.80
049	0.25	0.15	0.07	—	0.23	—	—	—	0.05	0.20
050	2.12	1.27	0.59	—	1.95	—	—	—	0.42	1.70
051	8.80	7.14	2.81	0.31	4.81	4.90	5.46	3.91	1.08	6.71
052	9.71	7.88	3.10	0.34	5.31	5.41	6.03	4.32	1.19	7.40
053	8.62	7.31	2.80	0.15	4.81	5.17	5.52	4.00	0.98	5.82
054	9.37	7.95	3.04	0.16	5.23	5.62	6.00	4.35	1.07	6.33
055	0.31	0.07	0.03	0.01	0.03	0.01	0.04	0.03	—	0.04
056	0.35	0.08	0.03	0.01	0.03	0.01	0.05	0.03	—	0.05
057	0.58	0.55	0.13	0.14	0.31	0.22	—	0.30	0.06	0.40
058	2.74	2.59	0.61	0.66	1.46	1.04	—	1.42	0.28	1.89
059	0.98	0.49	0.13	0.22	0.53	0.18	0.40	0.43	—	0.51
060	1.40	0.52	0.14	0.24	0.56	0.19	0.43	0.45	—	0.54
061	1.11	0.36	0.16	0.14	0.71	0.71	—	0.50	0.20	0.71
062	1.24	0.40	0.18	0.16	0.79	0.79	—	0.56	0.22	0.79
063	0.98	0.48	0.15	—	0.64	0.32	0.57	0.46	—	0.63
064	1.08	0.53	0.16	—	0.71	0.35	0.63	0.51	—	0.70
065	0.90	0.42	0.19	—	0.58	0.58	0.67	0.44	—	0.58
066	0.98	0.46	0.21	—	0.63	0.63	0.73	0.48	—	0.63
067	0.81	0.55	0.24	0.20	0.42	0.24	—	0.45	0.10	0.60
068	2.46	1.66	0.74	0.62	1.26	0.74	—	1.35	0.31	1.82
069	—	0.20	0.08	0.11	—	—	—	—	0.06	—
070	—	0.22	0.09	0.12	—	—	—	—	0.07	—
071	—	0.20	0.09	0.11	—	—	—	—	0.06	—
072	—	0.22	0.10	0.12	—	—	—	—	0.07	—
073	1.88	1.08	0.27	0.36	1.08	0.63	0.90	1.17	0.45	1.17
074	2.10	1.20	0.30	0.40	1.20	0.70	1.00	1.30	0.50	1.30

TABLE 4   Composition of Important Feeds: Amino Acids—*Continued*

Entry Number	Feed Name Description	International Feed Number	Dry Matter (%)	Crude Protein (%)	Arginine (%)	Glycine (%)	Histidine (%)	Isoleucine (%)
	COCONUT *Cocos nucifera*							
075	meats, meal mech extd (Copra	5-01-572	92.0	20.7	2.40	1.05	0.42	0.63
076	meal)		100.0	22.4	2.60	1.13	0.46	0.68
077	meats, meal solv extd (Copra	5-01-573	91.0	21.3	2.41	1.03	0.38	0.83
078	meal)		100.0	23.4	2.65	1.14	0.41	0.91
	COFFEE *Coffea* spp							
079	fruit without seeds, dehy (Coffee	1-09-734	87.0	11.8	0.55	0.75	0.44	0.47
080	pulp)		100.0	13.6	0.63	0.86	0.50	0.54
	CORN *Zea mays* grain—see Corn grain, dent white, dent yellow, or flint							
	CORN, DENT YELLOW *Zea mays indentata*							
081	distillers grains, dehy	5-28-235	94.0	27.9	0.97	0.49	0.62	0.99
082			100.0	29.8	1.04	0.52	0.67	1.06
083	distillers grains with solubles,	5-28-236	92.0	27.1	0.96	0.51	0.64	1.39
084	dehy		100.0	29.5	1.05	0.55	0.70	1.52
085	distillers solubles, dehy	5-28-237	93.0	27.6	0.97	1.11	0.68	1.33
086			100.0	29.7	1.05	1.20	0.73	1.43
087	ears, ground (Corn and cob	4-28-238	87.0	7.8	0.36	0.31	0.16	0.35
088	meal)		100.0	9.0	0.42	0.36	0.19	0.40
089	germs, meal wet milled solv extd	5-28-240	91.0	20.4	1.30	1.10	0.69	0.69
090			100.0	22.3	1.43	1.20	0.76	0.76
091	gluten, meal	5-28-241	91.0	42.7	1.39	1.51	0.97	2.25
092			100.0	46.8	1.53	1.65	1.06	2.46
093	gluten, meal, 60% protein	5-28-242	90.0	60.7	2.08	2.10	1.40	2.54
094			100.0	67.2	2.31	2.33	1.55	2.82
095	gluten with bran (Corn gluten	5-28-243	90.0	23.0	0.78	0.85	0.61	0.88
096	feed)		100.0	25.6	0.87	0.94	0.68	0.98
097	grain	4-02-935	89.0	9.6	0.43	0.37	0.26	0.35
098			100.0	10.9	0.48	0.42	0.29	0.39
099	grain, ground	4-26-023	88.0	8.8	0.47	0.33	0.22	0.34
100			100.0	10.0	0.54	0.38	0.25	0.39
101	grain, flaked	4-28-244	89.0	9.9	0.44	0.36	0.28	0.34
102			100.0	11.2	0.49	0.40	0.31	0.38
103	grain, opaque 2 (High lysine)	4-28-253	90.0	10.1	0.66	0.48	0.35	0.35
104			100.0	11.3	0.73	0.53	0.39	0.38
105	grits by-product (Hominy feed)	4-03-011	90.0	10.4	0.47	0.34	0.19	0.39
106			100.0	11.5	0.52	0.38	0.22	0.43
107	silage	3-02-912	30.0	2.5	0.29	0.11	0.06	0.08
108			100.0	8.3	0.97	0.36	0.21	0.25
	CORN, DENT WHITE *Zea mays indentata*							
109	grits by-product (Hominy feed)	4-02-990	90.0	10.6	0.43	0.27	0.19	0.34
110			100.0	11.8	0.48	0.30	0.21	0.37
	CORN, FLINT *Zea mays indurata*							
111	grain	4-02-948	89.0	9.9	—	—	—	—
112			100.0	11.1	—	—	—	—
	COTTON *Gossypium* spp							
113	seeds, meal mech extd (Whole	5-01-609	93.0	37.9	4.48	—	0.91	1.65
114	pressed cottonseed)		100.0	40.8	4.82	—	0.98	1.77

Entry Number	Leucine (%)	Lysine (%)	Methionine (%)	Cystine (%)	Phenylalanine (%)	Tyrosine (%)	Serine (%)	Threonine (%)	Tryptophan (%)	Valine (%)
075	1.26	0.59	0.32	0.21	0.84	0.52	—	0.61	0.20	0.94
076	1.36	0.64	0.34	0.23	0.91	0.57	—	0.66	0.22	1.02
077	1.44	0.60	0.32	0.25	0.86	0.57	—	0.66	0.20	1.04
078	1.59	0.66	0.35	0.27	0.95	0.63	—	0.73	0.22	1.14
079	0.86	0.76	0.15	0.11	0.55	0.40	0.71	0.52	—	0.83
080	0.98	0.87	0.17	0.13	0.63	0.46	0.81	0.60	—	0.95
081	3.01	0.78	0.40	0.24	0.94	0.84	—	0.49	0.20	1.18
082	3.22	0.84	0.43	0.26	1.00	0.90	—	0.52	0.21	1.26
083	2.23	0.70	0.50	0.29	1.51	0.70	1.30	0.93	0.17	1.50
084	2.43	0.77	0.54	0.32	1.64	0.76	1.42	1.01	0.19	1.63
085	2.36	0.91	0.56	0.45	1.49	0.87	1.22	1.02	0.24	1.55
086	2.54	0.99	0.60	0.48	1.60	0.94	1.32	1.10	0.26	1.67
087	0.86	0.17	0.14	0.12	0.39	0.32	—	0.28	0.07	0.31
088	1.00	0.20	0.16	0.14	0.45	0.38	—	0.33	0.08	0.36
089	1.79	0.90	0.58	0.40	0.90	0.69	1.00	1.09	0.20	1.19
090	1.97	0.98	0.64	0.44	0.98	0.76	1.09	1.19	0.21	1.31
091	7.22	0.80	1.04	0.67	2.78	1.01	1.80	1.42	0.21	2.19
092	7.92	0.87	1.14	0.73	3.05	1.11	1.97	1.56	0.23	2.40
093	10.23	1.01	1.78	0.99	4.02	3.19	3.35	2.22	0.30	3.09
094	11.33	1.12	1.98	1.10	4.45	3.54	3.71	2.46	0.33	3.43
095	2.20	0.64	0.37	0.44	0.81	0.72	0.85	0.78	0.15	1.10
096	2.44	0.71	0.41	0.49	0.90	0.81	0.94	0.87	0.17	1.22
097	1.21	0.25	0.17	0.22	0.48	0.38	0.50	0.35	0.08	0.44
098	1.37	0.28	0.19	0.25	0.54	0.43	0.57	0.40	0.09	0.50
099	0.99	0.21	0.18	0.16	0.43	0.38	0.46	0.35	0.08	0.44
100	1.12	0.24	0.21	0.18	0.49	0.43	0.53	0.39	0.09	0.51
101	1.24	0.25	0.15	0.25	0.44	0.39	0.48	0.35	—	0.47
102	1.40	0.28	0.17	0.28	0.50	0.44	0.54	0.39	—	0.53
103	0.99	0.42	0.17	0.20	0.43	0.40	0.47	0.37	0.11	0.50
104	1.10	0.46	0.19	0.22	0.48	0.44	0.52	0.41	0.12	0.56
105	0.85	0.38	0.16	0.15	0.33	0.50	—	0.39	0.11	0.49
106	0.94	0.42	0.18	0.16	0.36	0.55	—	0.44	0.12	0.55
107	0.28	0.13	0.13	—	0.11	0.06	0.12	0.11	—	0.14
108	0.93	0.43	0.44	—	0.36	0.20	0.40	0.36	—	0.45
109	0.82	0.36	0.12	0.11	0.34	0.40	—	0.34	0.11	0.44
110	0.91	0.40	0.13	0.12	0.37	0.44	—	0.37	0.12	0.48
111	—	0.27	0.18	—	—	—	—	—	0.09	—
112	—	0.30	0.20	—	—	—	—	—	0.10	—
113	2.53	1.91	0.73	0.64	2.50	1.35	—	1.57	0.66	2.08
114	2.72	2.06	0.78	0.69	2.69	1.45	—	1.69	0.71	2.24

TABLE 4 Composition of Important Feeds: Amino Acids—*Continued*

Entry Number	Feed Name Description	International Feed Number	Dry Matter (%)	Crude Protein (%)	Arginine (%)	Glycine (%)	Histidine (%)	Isoleucine (%)
115	seeds, meal mech extd, 36%	5-01-625	92.0	38.6	3.56	1.83	0.92	1.32
116	protein		100.0	41.9	3.86	1.99	0.99	1.44
117	seeds, meal mech extd, 41%	5-01-617	93.0	41.0	4.18	1.91	1.07	1.45
118	protein		100.0	44.3	4.51	2.06	1.15	1.56
119	seeds, meal prepressed solv	5-07-872	91.0	41.3	4.27	1.94	1.15	1.44
120	extd, 41% protein		100.0	45.6	4.71	2.14	1.27	1.59
121	seeds, meal prepressed solv	5-07-873	91.0	44.7	4.77	1.80	1.48	1.36
122	extd, 44% protein		100.0	48.9	5.22	1.97	1.62	1.49
123	seeds, meal solv extd, low	5-01-633	93.0	41.5	—	—	—	—
124	gossypol		100.0	44.8	—	—	—	—
125	seeds, meal solv extd, 41%	5-01-621	91.0	41.2	4.21	1.98	1.11	1.52
126	protein		100.0	45.2	4.62	2.17	1.21	1.67
127	seeds without hulls, meal pre-	5-07-874	93.0	50.3	4.83	2.82	1.21	1.48
128	pressed solv extd, 50% protein		100.0	54.0	5.20	3.03	1.30	1.59
	COWPEA, COMMON *Vigna sinensis*							
129	hay, sun-cured	1-01-645	90.0	17.5	1.11	—	0.45	1.26
130			100.0	19.4	1.23	—	0.50	1.40
	CRAB *Callinectes sapidus–Cancer* spp							
131	process residue, meal (Crab	5-01-663	92.0	32.1	1.66	1.75	0.49	1.17
132	meal)		100.0	34.8	1.80	1.89	0.53	1.26
	DISTILLERS GRAINS—SEE CORN, SEE SORGHUM							
	EMMER *Triticum dicoccum*							
133	grain	4-01-830	91.0	11.7	0.46	—	0.20	0.42
134			100.0	12.9	0.51	—	0.22	0.46
	FISH							
135	solubles, condensed	5-01-969	50.0	32.7	1.63	3.85	1.43	1.03
136			100.0	65.3	3.25	7.68	2.85	2.06
137	solubles, dehy	5-01-971	93.0	64.1	3.05	5.74	2.10	2.05
138			100.0	69.2	3.29	6.20	2.26	2.21
	FISH, ALEWIFE *Pomolobus pseudoharengus*							
139	meal mech extd	5-09-830	90.0	36.4	5.37	4.24	2.19	3.90
140			100.0	40.6	5.98	4.72	2.44	4.34
	FISH, ANCHOVY *Engraulis ringen*							
141	meal mech extd	5-01-985	92.0	65.5	3.77	3.69	1.61	3.10
142			100.0	71.2	4.11	4.01	1.76	3.38
	FISH, CARP *Cyprinus carpio*							
143	meal mech extd	5-01-987	90.0	52.7	—	—	—	—
144			100.0	58.6	—	—	—	—
	FISH, HERRING *Clupea harengus*							
145	meal mech extd	5-02-000	92.0	72.0	4.62	4.41	1.65	3.13
146			100.0	78.3	5.02	4.80	1.80	3.41
	FISH, MENHADEN *Brevoortia tyrannus*							
147	meal mech extd	5-02-009	92.0	61.1	3.75	4.19	1.45	2.88
148			100.0	66.7	4.09	4.57	1.58	3.15

Entry Number	Leucine (%)	Lysine (%)	Methio-nine (%)	Cystine (%)	Phenyl-alanine (%)	Tyrosine (%)	Serine (%)	Threo-nine (%)	Trypto-phan (%)	Valine (%)
115	—	1.22	0.55	0.79	1.88	—	—	1.12	0.46	2.85
116	—	1.33	0.60	0.86	2.04	—	—	1.21	0.50	3.09
117	2.32	1.60	0.58	0.73	2.18	0.94	1.71	1.34	0.53	1.90
118	2.50	1.73	0.62	0.78	2.35	1.01	1.84	1.44	0.57	2.05
119	2.42	1.82	0.56	0.81	2.01	1.15	1.82	1.34	0.51	1.99
120	2.67	2.01	0.62	0.90	2.21	1.27	2.01	1.48	0.56	2.20
121	2.44	1.73	0.61	1.12	1.55	1.45	2.17	1.49	0.55	1.91
122	2.67	1.89	0.67	1.22	1.70	1.59	2.37	1.63	0.60	2.10
123	—	1.72	—	—	—	—	—	—	—	—
124	—	1.85	—	—	—	—	—	—	—	—
125	2.33	1.69	0.59	0.77	2.24	1.03	1.75	1.38	0.56	1.88
126	2.56	1.86	0.64	0.85	2.46	1.13	1.92	1.52	0.61	2.06
127	2.28	1.70	0.76	1.05	2.62	0.81	—	1.66	0.62	2.16
128	2.45	1.82	0.81	1.13	2.81	0.87	—	1.78	0.67	2.32
129	2.00	1.08	0.51	—	1.25	—	—	1.06	0.52	1.43
130	2.22	1.20	0.57	—	1.39	—	—	1.18	0.58	1.59
131	1.54	1.38	0.53	0.24	1.16	1.17	1.38	1.00	0.29	1.47
132	1.67	1.50	0.57	0.26	1.26	1.26	1.50	1.09	0.32	1.59
133	0.67	0.29	0.16	—	0.46	—	—	0.38	0.12	0.47
134	0.74	0.32	0.18	—	0.51	—	—	0.42	0.13	0.52
135	1.86	1.86	0.71	0.27	1.02	0.44	1.03	0.87	0.34	1.22
136	3.72	3.71	1.42	0.54	2.04	0.87	2.05	1.73	0.68	2.43
137	2.97	3.51	1.18	0.62	1.53	0.85	2.03	1.35	0.59	2.10
138	3.21	3.79	1.27	0.66	1.65	0.92	2.19	1.46	0.64	2.26
139	6.20	6.29	2.19	0.56	3.32	3.10	—	3.77	0.70	4.12
140	6.90	7.00	2.44	0.62	3.70	3.45	—	4.20	0.78	4.58
141	4.99	5.04	1.99	0.60	2.78	2.24	2.41	2.76	0.75	3.50
142	5.43	5.49	2.16	0.66	3.03	2.44	2.63	3.00	0.82	3.87
143	—	—	1.40	—	—	—	—	—	—	—
144	—	—	1.56	—	—	—	—	—	—	—
145	5.19	5.36	2.08	0.74	2.71	2.20	2.65	2.90	0.77	4.30
146	5.64	5.83	2.27	0.81	2.94	2.39	2.88	3.16	0.83	4.68
147	4.48	4.72	1.75	0.56	2.46	1.94	2.23	2.50	0.65	3.22
148	4.89	5.15	1.91	0.61	2.69	2.12	2.43	2.73	0.71	3.52

TABLE 4   Composition of Important Feeds: Amino Acids—*Continued*

Entry Number	Feed Name Description	International Feed Number	Dry Matter (%)	Crude Protein (%)	Arginine (%)	Glycine (%)	Histidine (%)	Isoleucine (%)
	FISH, REDFISH *Sciaenops ocellata*							
149	meal mech extd	5-07-973	93.0	56.8	4.06	4.05	1.30	3.46
150			100.0	61.0	4.36	4.35	1.39	3.72
	FISH, SALMON *Oncorhynchus* spp–*Salmo* spp							
151	meal mech extd	5-02-012	93.0	61.1	5.20	5.20	—	—
152			100.0	65.6	5.59	5.59	—	—
	FISH, SARDINE *Clupea* spp–*Sardinops* spp							
153	meal mech extd	5-02-015	93.0	65.2	2.70	4.50	1.80	3.34
154			100.0	70.0	2.90	4.84	1.93	3.59
	FISH, TUNA *Thunnus thynnus–Thunnus albacares*							
155	meal mech extd	5-02-023	93.0	59.0	3.43	4.09	1.75	2.45
156			100.0	63.6	3.69	4.41	1.89	2.64
157	process residue, meal dehy	5-07-977	94.0	53.4	3.43	4.08	1.47	2.38
158			100.0	56.8	3.65	4.34	1.56	2.53
	FISH, TURBOT *Psetta maxima*							
159	whole, fresh	5-07-978	25.0	14.4	—	—	—	—
160			100.0	57.3	—	—	—	—
	FISH, WHITE *Gadidae* (family)–*Lophiidae* (family)							
161	meal mech extd	5-02-025	91.0	62.2	4.02	4.42	1.34	2.72
162			100.0	68.2	4.41	4.84	1.47	2.98
	FLAX *Linum usitatissimum*							
163	seeds, meal mech extd (Linseed	5-02-045	91.0	34.3	2.81	1.63	0.65	1.69
164	meal)		100.0	37.9	3.10	1.80	0.71	1.86
165	seeds, meal solv extd (Linseed	5-02-048	90.0	34.6	2.94	1.74	0.69	1.68
166	meal)		100.0	38.3	3.25	1.93	0.77	1.87
	GELATIN							
167	process residue (Gelatin by-	5-14-503	90.0	87.6	6.97	19.30	0.76	1.38
168	products)		100.0	97.4	7.75	21.48	0.85	1.54
	GRAPE *Vitis* spp							
169	marc, dehy (Pomace)	1-02-208	91.0	11.8	0.67	0.90	0.26	0.55
170			100.0	13.0	0.74	0.99	0.29	0.60
	GROUNDNUT—SEE PEANUT							
	HOG MILLET—SEE MILLET, PROSO							
	HOMINY FEED—SEE CORN							
	LINSEED—SEE FLAX							
	LIVERS							
171	meal	5-00-389	92.0	66.0	4.04	5.60	1.48	3.10
172			100.0	71.4	4.37	6.05	1.60	3.36

Entry Number	Leucine (%)	Lysine (%)	Methionine (%)	Cystine (%)	Phenylalanine (%)	Tyrosine (%)	Serine (%)	Threonine (%)	Tryptophan (%)	Valine (%)
149	4.86	6.56	1.80	0.41	2.50	1.68	—	2.60	0.60	3.30
150	5.22	7.04	1.94	0.44	2.68	1.81	—	2.79	0.65	3.55
151	—	7.60	1.60	0.70	—	—	—	—	0.50	—
152	—	8.17	1.72	0.75	—	—	—	—	0.54	—
153	—	5.91	2.01	0.80	2.00	2.80	—	2.60	0.50	4.10
154	—	6.34	2.16	0.86	2.15	3.00	—	2.79	0.54	4.40
155	3.79	4.22	1.47	0.47	2.15	1.69	2.09	2.31	0.57	2.77
156	4.09	4.54	1.58	0.50	2.32	1.82	2.25	2.49	0.62	2.98
157	3.85	3.89	1.47	0.42	2.19	2.04	2.28	2.31	0.56	2.83
158	4.09	4.14	1.56	0.45	2.33	2.17	2.43	2.46	0.60	3.01
159	—	—	—	—	—	0.11	0.17	0.10	—	—
160	—	—	—	—	—	0.44	0.67	0.41	—	—
161	4.36	4.53	1.68	0.75	2.28	1.83	3.06	2.57	0.67	3.02
162	4.78	4.96	1.84	0.82	2.50	2.00	3.35	2.82	0.73	3.31
163	1.92	1.18	0.58	0.61	1.38	0.96	1.89	1.14	0.50	1.61
164	2.11	1.30	0.64	0.67	1.53	1.06	2.09	1.25	0.56	1.77
165	2.02	1.16	0.54	0.61	1.46	1.09	1.92	1.22	0.51	1.74
166	2.24	1.28	0.60	0.67	1.62	1.21	2.13	1.35	0.56	1.93
167	2.91	3.55	0.73	0.13	1.79	0.52	3.10	1.76	0.05	2.09
168	3.24	3.95	0.81	0.15	1.99	0.58	3.45	1.96	0.05	2.33
169	1.64	0.50	0.18	0.17	0.55	0.16	—	0.38	0.07	1.10
170	1.80	0.55	0.20	0.19	0.60	0.18	—	0.42	0.08	1.21
171	5.31	5.21	1.22	0.94	2.92	1.70	2.50	2.49	0.69	4.15
172	5.74	5.63	1.32	1.01	3.15	1.84	2.70	2.70	0.74	4.49

TABLE 4 Composition of Important Feeds: Amino Acids—*Continued*

Entry Number	Feed Name Description	International Feed Number	Dry Matter (%)	Crude Protein (%)	Arginine (%)	Glycine (%)	Histidine (%)	Isoleucine (%)
	MAIZE—SEE CORN							
	MANURE—SEE CATTLE, SEE POULTRY							
	MEAT							
173	meal rendered	5-00-385	94.0	51.4	3.60	6.29	0.96	1.75
174			100.0	54.8	3.84	6.71	1.02	1.86
175	with blood, meal rendered	5-00-386	92.0	59.4	3.59	6.61	1.83	1.93
176	(Tankage)		100.0	64.5	3.90	7.17	1.99	2.09
177	with blood with bone, meal rendered (Tankage)	5-00-387	93.0	46.6	2.82	6.58	1.76	1.87
178			100.0	50.2	3.03	7.08	1.90	2.01
179	with bone, meal rendered	5-00-388	93.0	50.4	3.49	6.45	0.96	1.64
180			100.0	54.1	3.75	6.93	1.04	1.76
	MILK							
181	dehy (Cattle)	5-01-167	96.0	25.4	0.92	—	0.72	1.33
182			100.0	26.5	0.96	—	0.75	1.39
183	fresh (Cattle)	5-01-168	12.0	3.3	—	—	—	0.32
184			100.0	26.7	—	—	—	2.58
185	skimmed dehy (Cattle)	5-01-175	94.0	33.7	1.15	0.29	0.86	2.18
186			100.0	35.8	1.23	0.31	0.92	2.32
187	skimmed fresh (Cattle)	5-01-170	10.0	3.0	—	—	—	—
188			100.0	31.2	—	—	—	—
	MILLET, PEARL—SEE PEARLMILLET							
	MILLET, PROSO *Panicum miliaceum*							
189	grain	4-03-120	90.0	11.6	0.36	—	0.21	0.45
190			100.0	12.9	0.39	—	0.23	0.50
	OATS *Avena sativa*							
191	breakfast cereal by-product, less than 4% fiber (Feeding oat meal) (Oat middlings)	4-03-303	91.0	14.8	0.83	0.62	0.30	0.54
192			100.0	16.4	0.92	0.69	0.33	0.60
193	grain	4-03-309	89.0	11.8	0.70	0.46	0.18	0.43
194			100.0	13.3	0.79	0.52	0.21	0.49
195	grain, Pacific Coast	4-07-999	91.0	9.1	0.58	0.40	0.17	0.38
196			100.0	10.0	0.63	0.44	0.18	0.42
197	groats	4-03-331	90.0	15.8	0.86	0.60	0.25	0.55
198			100.0	17.7	0.96	0.67	0.28	0.61
199	hulls	1-03-281	92.0	3.6	0.17	0.14	0.09	0.17
200			100.0	3.9	0.19	0.15	0.09	0.19
	PEA *Pisum* spp							
201	seeds	5-03-600	89.0	22.5	1.39	1.08	0.65	1.14
202			100.0	25.3	1.56	1.22	0.73	1.28
	PEANUT *Arachis hypogaea*							
203	kernels, meal mech extd (Peanut meal)	5-03-649	93.0	48.1	5.06	2.40	1.08	1.69
204			100.0	52.0	5.46	2.59	1.17	1.83
205	kernels, meal solv extd (Peanut meal)	5-03-650	92.0	48.1	4.55	2.35	0.95	1.76
206			100.0	52.3	4.95	2.56	1.03	1.91
	PEARLMILLET *Pennisetum glaucum*							
207	grain	2-03-118	90.0	15.7	0.74	0.47	0.31	0.37
208			100.0	17.5	0.82	0.52	0.34	0.41

Entry Number	Leucine (%)	Lysine (%)	Methionine (%)	Cystine (%)	Phenylalanine (%)	Tyrosine (%)	Serine (%)	Threonine (%)	Tryptophan (%)	Valine (%)
173	3.19	3.23	0.70	0.65	1.81	0.96	2.16	1.64	0.34	2.52
174	3.40	3.45	0.75	0.70	1.94	1.02	2.30	1.75	0.37	2.68
175	5.12	3.74	0.73	0.45	2.54	1.29	—	2.32	0.65	3.77
176	5.56	4.06	0.79	0.49	2.76	1.40	—	2.52	0.70	4.10
177	5.27	3.32	0.69	0.27	2.28	—	—	2.18	0.62	3.42
178	5.67	3.57	0.74	0.29	2.46	—	—	2.35	0.67	3.68
179	3.06	2.90	0.65	0.50	1.70	0.79	1.81	1.65	0.30	2.45
180	3.29	3.11	0.70	0.53	1.83	0.85	1.94	1.77	0.32	2.63
181	2.56	2.25	0.61	—	1.33	1.33	—	1.02	0.41	1.74
182	2.67	2.35	0.64	—	1.39	1.39	—	1.07	0.43	1.81
183	0.25	0.28	0.09	—	0.16	—	—	0.16	0.05	0.25
184	2.03	2.27	0.69	—	1.33	—	—	1.33	0.39	2.03
185	3.32	2.53	0.90	0.45	1.56	1.14	1.67	1.56	0.43	2.28
186	3.53	2.70	0.96	0.48	1.66	1.22	1.78	1.67	0.46	2.43
187	—	0.28	—	—	—	—	—	—	—	—
188	—	2.92	—	—	—	—	—	—	—	—
189	1.15	0.26	0.29	—	0.57	—	—	0.40	0.17	0.58
190	1.28	0.29	0.32	—	0.63	—	—	0.44	0.19	0.64
191	1.06	0.54	0.21	0.25	0.69	0.72	0.70	0.48	0.20	0.73
192	1.17	0.59	0.23	0.28	0.76	0.79	0.77	0.53	0.22	0.80
193	0.81	0.39	0.17	0.19	0.52	0.46	0.44	0.36	0.15	0.56
194	0.91	0.44	0.19	0.22	0.58	0.52	0.50	0.40	0.17	0.63
195	0.70	0.33	0.13	0.17	0.43	0.70	0.40	0.30	0.12	0.49
196	0.77	0.37	0.14	0.18	0.47	0.77	0.44	0.33	0.13	0.54
197	1.04	0.53	0.20	0.20	0.67	0.57	—	0.45	0.19	0.76
198	1.16	0.59	0.23	0.23	0.75	0.64	—	0.50	0.21	0.84
199	0.28	0.17	0.09	0.06	0.17	0.17	—	0.17	0.09	0.20
200	0.30	0.19	0.09	0.07	0.18	0.19	—	0.18	0.09	0.22
201	1.78	1.54	0.28	0.19	1.25	—	—	0.93	0.22	1.25
202	2.00	1.73	0.32	0.22	1.41	—	—	1.04	0.25	1.41
203	3.02	1.50	0.49	0.75	2.34	1.66	1.44	1.24	0.47	2.08
204	3.26	1.62	0.53	0.81	2.53	1.79	1.56	1.34	0.51	2.24
205	2.70	1.77	0.42	0.73	2.04	1.51	3.10	1.16	0.48	1.88
206	2.94	1.93	0.46	0.79	2.22	1.65	3.37	1.26	0.52	2.04
207	1.14	0.45	0.25	—	0.56	0.35	0.74	0.48	—	0.49
208	1.27	0.50	0.28	—	0.63	0.39	0.82	0.54	—	0.55

TABLE 4   Composition of Important Feeds: Amino Acids—*Continued*

Entry Number	Feed Name Description	International Feed Number	Dry Matter (%)	Crude Protein (%)	Arginine (%)	Glycine (%)	Histidine (%)	Isoleucine (%)
	POTATO *Solanum tuberosum*							
209	tubers, dehy	4-07-850	91.0	8.1	0.26	—	0.15	0.25
210			100.0	8.9	0.28	—	0.17	0.28
	POULTRY							
211	by-product, meal rendered	5-03-798	93.0	58.7	3.77	5.42	1.01	2.38
212	(Viscera with feet with heads)		100.0	62.8	4.03	5.80	1.08	2.54
213	feathers, hydrolyzed	5-03-795	93.0	84.9	7.05	6.44	0.99	4.06
214			100.0	91.3	7.58	6.92	1.06	4.37
215	manure with litter, dehy	5-05-587	89.0	21.9	0.40	1.60	0.20	0.46
216			100.0	24.5	0.45	1.79	0.22	0.51
217	manure without litter, dehy	5-14-015	90.0	25.5	0.46	1.02	0.20	0.53
218			100.0	28.2	0.51	1.13	0.22	0.58
	RAPE *Brassica* spp							
219	seeds, meal mech extd	5-03-870	92.0	35.6	1.99	1.88	0.90	1.41
220			100.0	38.7	2.16	2.04	0.98	1.53
221	seeds, meal solv extd	5-03-871	91.0	37.0	2.06	1.79	0.99	1.35
222			100.0	40.6	2.26	1.97	1.09	1.48
	RAPE, SUMMER *Brassica napus*							
223	seeds, meal mech extd	5-08-136	94.0	35.2	1.79	1.65	0.85	1.31
224			100.0	37.4	1.91	1.75	0.90	1.39
225	seeds, meal prepressed solv extd	5-08-135	92.0	40.5	2.23	1.94	1.09	1.46
226			100.0	44.0	2.42	2.11	1.19	1.59
	RICE *Oryza sativa*							
227	bran with germ (Rice bran)	4-03-928	91.0	12.7	0.72	0.80	0.23	0.46
228			100.0	14.1	0.79	0.88	0.25	0.51
229	grain, ground (Ground rough	4-03-938	89.0	7.9	0.57	0.55	0.14	0.31
230	rice) (Ground paddy rice)		100.0	8.9	0.64	0.62	0.15	0.34
231	grain, polished and broken	4-03-932	89.0	7.6	0.49	0.34	0.18	0.33
232	(Brewers rice)		100.0	8.6	0.56	0.38	0.20	0.37
233	groats, polished (Rice, polished)	4-03-942	89.0	7.2	0.44	0.74	0.18	0.45
234			100.0	8.2	0.50	0.83	0.20	0.50
235	polishings	4-03-943	90.0	12.1	0.51	0.70	0.17	0.35
236			100.0	13.4	0.57	0.78	0.19	0.39
	RYE *Secale cereale*							
237	fresh	2-04-018	24.0	3.8	—	—	—	0.41
238			100.0	15.9	—	—	—	1.70
239	grain	4-04-047	88.0	12.1	0.53	0.49	0.26	0.47
240			100.0	13.8	0.61	0.56	0.29	0.53
	RYEGRASS, ITALIAN *Lolium multiflorum*							
241	fresh	2-04-073	25.0	3.5	—	—	—	0.39
242			100.0	14.5	—	—	—	1.60
	SAFFLOWER *Carthamus tinctorius*							
243	seeds, meal mech extd	5-04-109	91.0	20.2	1.38	1.11	0.44	0.56
244			100.0	22.1	1.51	1.21	0.48	0.62
245	seeds, meal solv extd	5-04-110	92.0	23.4	1.95	1.13	—	0.28
246			100.0	25.4	2.11	1.22	—	0.30
247	seeds without hulls, meal solv extd	5-07-959	92.0	43.0	3.65	2.32	1.07	1.56
248			100.0	46.9	3.98	2.54	1.16	1.70

SCREENINGS—SEE BARLEY,
SEE CEREALS, SEE WHEAT

Entry Number	Leucine (%)	Lysine (%)	Methio-nine (%)	Cystine (%)	Phenyl-alanine (%)	Tyrosine (%)	Serine (%)	Threo-nine (%)	Trypto-phan (%)	Valine (%)
209	0.60	0.41	0.10	0.07	0.40	—	—	0.47	0.14	0.36
210	0.66	0.45	0.11	0.08	0.44	—	—	0.52	0.15	0.40
211	4.00	2.89	1.06	0.92	1.84	0.94	2.62	1.94	0.46	2.86
212	4.28	3.10	1.13	0.98	1.97	1.01	2.81	2.08	0.50	3.06
213	6.94	2.32	0.55	3.24	3.05	2.32	9.26	3.97	0.52	6.48
214	7.46	2.49	0.59	3.48	3.28	2.49	9.96	4.27	0.56	6.97
215	0.79	0.44	0.12	0.13	0.41	0.27	0.50	0.43	0.54	0.59
216	0.88	0.49	0.13	0.14	0.46	0.30	0.56	0.48	0.60	0.66
217	0.78	0.48	0.16	0.77	0.46	0.30	0.54	0.50	0.52	0.71
218	0.86	0.53	0.18	0.86	0.51	0.33	0.60	0.56	0.57	0.79
219	2.41	1.67	0.68	0.30	1.42	0.85	1.45	1.53	0.48	1.81
220	2.62	1.82	0.74	0.32	1.54	0.92	1.58	1.66	0.53	1.97
221	2.50	1.98	0.71	0.30	1.41	0.79	1.57	1.56	0.43	1.79
222	2.74	2.18	0.78	0.33	1.55	0.87	1.72	1.72	0.47	1.96
223	2.27	1.55	0.66	—	1.32	0.76	1.42	1.44	0.33	1.68
224	2.42	1.64	0.70	—	1.40	0.81	1.51	1.53	0.35	1.78
225	2.71	2.15	0.77	—	1.54	0.85	1.70	1.70	0.49	1.94
226	2.95	2.33	0.84	—	1.67	0.92	1.85	1.85	0.53	2.11
227	0.70	0.49	0.23	0.10	0.44	0.69	0.77	0.43	0.10	0.69
228	0.77	0.54	0.26	0.11	0.49	0.76	0.85	0.47	0.11	0.76
229	0.56	0.27	0.16	0.12	0.33	0.54	0.45	0.24	0.11	0.44
230	0.63	0.30	0.18	0.13	0.37	0.60	0.50	0.27	0.12	0.50
231	0.68	0.27	0.12	0.08	0.39	0.41	0.41	0.24	0.10	0.47
232	0.77	0.30	0.14	0.09	0.44	0.46	0.46	0.27	0.11	0.53
233	0.71	0.28	0.25	0.09	0.53	0.62	—	0.36	0.09	0.53
234	0.80	0.32	0.28	0.11	0.60	0.70	—	0.40	0.11	0.60
235	0.70	0.52	0.20	0.13	0.38	0.42	—	0.34	0.10	0.72
236	0.78	0.58	0.22	0.14	0.42	0.46	—	0.38	0.11	0.80
237	0.63	0.34	0.07	—	0.26	—	—	0.55	0.05	0.34
238	2.60	1.40	0.30	—	1.10	—	—	2.30	0.20	1.40
239	0.70	0.42	0.17	0.19	0.56	0.26	0.52	0.36	0.11	0.56
240	0.80	0.48	0.19	0.21	0.64	0.30	0.60	0.41	0.13	0.64
241	0.57	0.37	0.07	—	0.30	—	—	0.52	0.07	0.32
242	2.30	1.50	0.30	—	1.20	—	—	2.10	0.30	1.30
243	1.21	0.68	0.39	0.81	1.16	—	—	0.56	0.29	1.09
244	1.33	0.74	0.42	0.88	1.26	—	—	0.61	0.32	1.19
245	—	0.72	0.34	0.36	—	—	—	0.51	0.27	—
246	—	0.78	0.37	0.39	—	—	—	0.56	0.29	—
247	2.46	1.27	0.68	0.70	1.75	1.07	—	1.30	0.59	2.33
248	2.68	1.38	0.74	0.76	1.91	1.17	—	1.42	0.65	2.54

TABLE 4 Composition of Important Feeds: Amino Acids—*Continued*

Entry Number	Feed Name Description	International Feed Number	Dry Matter (%)	Crude Protein (%)	Arginine (%)	Glycine (%)	Histidine (%)	Isoleucine (%)
	SESAME *Sesamum indicum*							
249	seeds, meal mech extd	5-04-220	93.0	45.5	4.55	3.97	1.07	1.96
250			100.0	49.1	4.91	4.28	1.16	2.12
	SHRIMP *Pandalus* spp–*Penaeus* spp							
251	process residue, meal (Shrimp	5-04-226	90.0	39.9	25.2	—	0.96	1.68
252	meal)		100.0	44.2	2.79	—	1.07	1.86
	SORGHUM *Sorghum bicolor*							
253	grain	4-04-383	90.0	11.1	0.39	0.34	0.23	0.45
254			100.0	12.4	0.43	0.38	0.26	0.50
255	grain, less than 8% protein	4-20-892	88.0	6.8	0.24	0.26	0.15	0.23
256			100.0	7.7	0.27	0.30	0.17	0.26
257	grain, 8–10% protein	4-20-893	87.0	8.8	0.34	0.35	0.19	0.42
258			100.0	10.1	0.39	0.41	0.22	0.49
259	grain, more than 10% protein	4-20-894	88.0	11.0	0.35	0.32	0.23	0.43
260			100.0	12.5	0.40	0.36	0.26	0.49
	SOYBEAN *Glycine max*							
261	flour by-product (Soybean mill	4-04-594	90.0	12.6	0.75	0.48	0.18	0.41
262	feed)		100.0	14.0	0.84	0.53	0.20	0.45
263	hulls	1-04-560	91.0	11.0	0.59	0.77	0.25	0.30
264			100.0	12.1	0.65	0.85	0.27	0.33
265	protein concentrate, more than	5-08-038	92.0	84.3	7.34	3.32	2.41	4.60
266	70% protein		100.0	91.9	8.00	3.62	2.63	5.02
267	seeds	5-04-610	92.0	39.2	2.85	1.52	0.97	2.12
268			100.0	42.8	3.11	1.66	1.06	2.32
269	seeds, heat processed	5-04-597	90.0	38.0	2.80	2.00	1.01	2.18
270			100.0	42.2	3.11	2.22	1.12	2.42
271	seeds, meal mech extd	5-04-600	90.0	42.9	3.07	2.38	1.14	2.63
272			100.0	47.7	3.41	2.64	1.26	2.92
273	seeds, meal solv extd	5-04-604	90.0	44.8	3.03	1.82	1.07	2.03
274			100.0	49.9	3.38	2.03	1.19	2.27
275	seeds, meal solv extd, 46%	5-26-146	90.0	47.0	3.39	2.44	1.29	2.47
276	protein		100.0	52.3	3.77	2.71	1.43	2.74
277	seeds without hulls, meal solv	5-04-612	90.0	49.7	3.67	2.42	1.22	2.46
278	extd		100.0	55.1	4.07	2.68	1.35	2.73
	SPELT *Triticum spelta*							
279	grain	4-04-651	90.0	12.0	0.45	—	0.18	0.36
280			100.0	13.3	0.50	—	0.20	0.40
	SUNFLOWER, COMMON *Helianthus annuus*							
281	seeds, meal solv extd	5-09-340	90.0	23.3	2.30	—	0.55	1.00
282			100.0	25.9	2.56	—	0.61	1.11
283	seeds without hulls, meal mech	5-04-738	93.0	41.4	3.45	1.77	0.90	1.76
284	extd		100.0	44.6	3.72	1.91	0.97	1.90
285	seeds without hulls, meal solv	5-04-739	93.0	46.3	4.42	2.82	1.23	2.25
286	extd		100.0	49.8	4.75	3.03	1.32	2.42
	TIMOTHY *Phleum pratense*							
287	hay, sun-cured	1-04-893	90.0	8.2	—	—	—	—
288			100.0	9.1	—	—	—	—
	TOMATO *Lycopersicon esculentum*							
289	pomace, dehy	5-05-041	92.0	21.6	1.20	—	0.40	0.70
290			100.0	23.5	1.30	—	0.43	0.76

Entry Number	Leucine (%)	Lysine (%)	Methio-nine (%)	Cystine (%)	Phenyl-alanine (%)	Tyrosine (%)	Serine (%)	Threo-nine (%)	Trypto-phan (%)	Valine (%)
249	3.20	1.27	1.37	0.59	2.14	1.87	2.95	1.60	0.71	2.33
250	3.45	1.36	1.48	0.64	2.31	2.02	3.18	1.72	0.76	2.51
251	2.68	2.17	0.82	0.59	1.59	1.33	—	1.42	0.36	1.83
252	2.98	2.41	0.91	0.66	1.76	1.47	—	1.58	0.40	2.03
253	1.44	0.25	0.13	0.20	0.56	0.41	0.50	0.36	0.11	0.52
254	1.60	0.28	0.15	0.22	0.62	0.46	0.55	0.40	0.12	0.58
255	0.74	0.18	0.08	0.08	0.32	0.15	0.30	0.22	—	0.35
256	0.84	0.20	0.09	0.09	0.36	0.17	0.34	0.25	—	0.40
257	1.18	0.21	0.16	0.16	0.42	0.38	—	0.29	0.10	0.53
258	1.35	0.24	0.18	0.19	0.48	0.44	—	0.33	0.12	0.61
259	1.37	0.22	0.15	0.11	0.52	0.17	—	0.33	—	0.54
260	1.56	0.25	0.17	0.12	0.59	0.19	—	0.38	—	0.61
261	0.58	0.65	0.13	0.14	0.38	0.23	—	0.30	0.13	0.38
262	0.64	0.72	0.14	0.16	0.42	0.26	—	0.34	0.14	0.42
263	0.71	0.64	0.12	0.07	0.41	0.39	0.65	1.29	0.07	0.36
264	0.78	0.70	0.13	0.08	0.45	0.43	0.71	1.42	0.08	0.40
265	6.33	5.61	0.88	0.92	4.33	3.10	5.19	3.34	0.88	4.38
266	6.90	6.12	0.96	1.00	4.71	3.38	5.66	3.64	0.96	4.77
267	3.00	2.44	0.54	0.55	2.03	1.02	2.14	1.66	0.54	2.06
268	3.28	2.67	0.59	0.60	2.22	1.12	2.33	1.81	0.59	2.25
269	—	2.40	0.54	0.55	2.10	—	—	1.69	0.52	2.02
270	—	2.67	0.60	0.61	2.33	—	—	1.88	0.58	2.24
271	3.62	2.79	0.65	0.56	2.20	1.55	2.01	1.72	0.61	2.28
272	4.02	3.10	0.72	0.63	2.45	1.72	2.23	1.92	0.68	2.53
273	3.27	2.68	0.52	0.75	2.11	1.33	2.11	1.66	0.64	2.02
274	3.65	2.99	0.58	0.83	2.36	1.48	2.36	1.85	0.71	2.25
275	3.84	3.05	0.60	0.70	2.50	1.49	—	1.98	0.58	2.46
276	4.27	3.39	0.67	0.78	2.78	1.66	—	2.20	0.64	2.73
277	3.73	3.17	0.71	0.75	2.44	1.68	—	1.94	0.69	2.55
278	4.14	3.52	0.79	0.83	2.71	1.86	—	2.15	0.77	2.82
279	0.63	0.27	0.18	—	0.45	—	—	0.36	0.09	0.45
280	0.70	0.30	0.20	—	0.50	—	—	0.40	0.10	0.50
281	1.60	1.00	0.50	0.50	1.15	—	1.00	1.05	0.45	1.60
282	1.78	1.11	0.56	0.56	1.28	—	1.11	1.17	0.50	1.78
283	2.47	1.61	0.94	0.69	1.80	1.00	—	1.37	0.50	2.01
284	2.66	1.73	1.01	0.74	1.94	1.08	—	1.47	0.54	2.17
285	3.83	1.92	1.16	0.74	2.36	1.39	2.20	1.93	0.61	2.60
286	4.12	2.06	1.25	0.79	2.54	1.49	2.37	2.07	0.65	2.80
287	—	—	—	—	—	2.40	—	—	—	—
288	—	—	—	—	—	2.66	—	—	—	—
289	1.70	1.60	0.10	—	0.90	0.90	—	0.70	0.20	1.00
290	1.85	1.74	0.11	—	0.98	0.98	—	0.76	0.22	1.09

TABLE 4   Composition of Important Feeds: Amino Acids—*Continued*

Entry Number	Feed Name Description	International Feed Number	Dry Matter (%)	Crude Protein (%)	Arginine (%)	Glycine (%)	Histidine (%)	Isoleucine (%)
	TORULA DRIED YEAST—SEE YEAST, TORULA							
	TRITICALE *Triticale hexaploide*							
291	grain	4-20-362	90.0	15.8	0.86	0.70	0.40	0.61
292			100.0	17.6	0.95	0.78	0.44	0.68
	WHALE *Balaena glacialis–Balaenoptera* spp							
293	meat, meal rendered	5-05-160	91.0	71.4	2.49	6.31	1.19	2.72
294			100.0	78.1	2.72	6.90	1.30	2.97
	WHEAT *Triticum aestivum*							
295	bran	4-05-190	89.0	15.2	0.96	0.86	0.39	0.51
296			100.0	17.1	1.09	0.97	0.44	0.57
297	bread, dehy	4-07-944	95.0	12.4	—	—	—	—
298			100.0	13.0	—	—	—	—
299	grain dust (Aspirated grain fines)	4-30-185	90.0	9.5	0.13	0.41	0.15	0.18
300			100.0	10.6	0.14	0.45	0.17	0.20
301	flour, hard red spring, less than	4-08-113	88.0	12.0	0.45	0.48	0.23	0.49
302	1.5% fiber		100.0	13.5	0.50	0.54	0.26	0.55
303	flour, less than 1.5% fiber	4-05-199	88.0	11.7	0.43	0.44	0.25	0.47
304	(Wheat feed flour)		100.0	13.4	0.49	0.51	0.28	0.53
305	flour, less than 2% fiber (Feed	4-28-221	88.0	11.0	0.41	0.43	0.23	0.44
306	flour)		100.0	12.5	0.46	0.49	0.26	0.50
307	flour by-product, less than 4%	4-05-203	88.0	15.3	0.96	0.74	0.41	0.55
308	fiber (Wheat red dog)		100.0	17.4	1.09	0.84	0.46	0.62
309	flour by-product, less than 4.5%	4-28-220	88.0	15.1	—	—	—	—
310	fiber (Middlings)		100.0	17.2	—	—	—	—
311	flour by-product, less than 7%	4-05-201	88.0	16.5	1.18	0.96	0.45	0.58
312	fiber (Wheat shorts)		100.0	18.6	1.34	1.09	0.51	0.66
313	flour by-product, less than 8%	4-28-219	88.0	16.0	—	—	—	—
314	fiber (Shorts)		100.0	18.2	—	—	—	—
315	flour by-product, less than 9.5%	4-05-205	89.0	16.4	0.92	0.51	0.38	0.67
316	fiber (Wheat middlings)		100.0	18.4	1.03	0.57	0.43	0.75
317	fresh	2-08-078	25.0	4.7	—	—	—	0.37
318			100.0	19.1	—	—	—	1.50
319	germs, ground	5-05-218	88.0	24.8	1.87	1.47	0.65	0.90
320			100.0	28.1	2.12	1.66	0.74	1.02
321	grain	4-05-211	89.0	14.2	0.59	0.57	0.29	0.47
322			100.0	16.0	0.67	0.65	0.32	0.53
323	grain, hard red spring	4-05-258	88.0	15.1	0.59	0.62	0.24	0.54
324			100.0	17.2	0.67	0.71	0.27	0.61
325	grain, hard red winter	4-05-268	88.0	12.7	0.64	0.57	0.30	0.51
326			100.0	14.4	0.73	0.65	0.34	0.58
327	grain screenings	4-05-216	89.0	14.1	0.40	0.83	0.30	0.46
328			100.0	15.8	0.44	0.93	0.34	0.52
329	grain, soft red winter	4-05-294	88.0	11.5	0.65	0.54	0.32	0.45
330			100.0	13.0	0.73	0.62	0.36	0.51
331	grain, soft white winter	4-05-337	89.0	10.1	0.46	0.49	0.22	0.41
332			100.0	11.3	0.52	0.55	0.24	0.46
333	grain, soft white winter, Pacific	4-08-555	89.0	10.0	0.45	0.50	0.20	0.40
334	Coast		100.0	11.2	0.50	0.56	0.22	0.45
335	mill run, less than 9.5% fiber	4-05-206	90.0	15.4	0.94	0.53	0.40	0.70
336			100.0	17.2	1.04	0.59	0.44	0.78
	WHEAT, DURUM *Triticum durum*							
337	grain	4-05-224	88.0	13.9	0.60	0.48	0.28	0.50
338			100.0	15.9	0.68	0.55	0.32	0.57

Entry Number	Leucine (%)	Lysine (%)	Methio-nine (%)	Cystine (%)	Phenyl-alanine (%)	Tyrosine (%)	Serine (%)	Threo-nine (%)	Trypto-phan (%)	Valine (%)
291	1.18	0.52	0.21	0.29	0.80	0.51	0.76	0.57	0.18	0.84
292	1.31	0.58	0.24	0.32	0.90	0.57	0.85	0.64	0.20	0.94
293	4.27	3.48	1.01	0.63	2.06	—	—	1.63	0.82	2.81
294	4.67	3.80	1.10	0.69	2.26	—	—	1.79	0.89	3.07
295	0.92	0.58	0.19	0.32	0.55	0.42	0.68	0.46	0.25	0.69
296	1.03	0.65	0.22	0.36	0.62	0.48	0.77	0.51	0.28	0.78
297	—	0.21	0.18	0.18	—	—	—	—	—	—
298	—	0.22	0.19	0.19	—	—	—	—	—	—
299	0.56	0.30	0.10	—	0.26	0.19	0.38	0.31	—	0.23
300	0.62	0.33	0.11	—	0.29	0.21	0.42	0.34	—	0.26
301	0.86	0.28	0.22	0.32	0.66	0.39	0.64	0.36	0.17	0.51
302	0.97	0.31	0.24	0.36	0.74	0.44	0.72	0.41	0.19	0.58
303	0.87	0.25	0.18	0.30	0.60	0.34	0.59	0.33	0.12	0.50
304	0.99	0.28	0.21	0.35	0.69	0.39	0.68	0.37	0.14	0.57
305	0.85	0.23	0.17	0.29	0.59	0.33	0.58	0.31	0.11	0.48
306	0.97	0.26	0.19	0.33	0.67	0.37	0.66	0.35	0.12	0.55
307	1.06	0.59	0.23	0.37	0.66	0.46	0.75	0.50	0.19	0.72
308	1.20	0.67	0.26	0.42	0.75	0.52	0.85	0.57	0.22	0.82
309	—	—	—	—	—	—	—	—	—	—
310	—	—	—	—	—	—	—	—	—	—
311	1.09	0.79	0.27	0.36	0.67	0.47	0.77	0.60	0.21	0.83
312	1.23	0.89	0.31	0.41	0.76	0.53	0.87	0.68	0.24	0.93
313	—	—	—	—	—	—	—	—	—	—
314	—	—	—	—	—	—	—	—	—	—
315	1.08	0.67	0.18	0.22	0.64	0.40	0.73	0.54	0.20	0.75
316	1.21	0.76	0.20	0.24	0.72	0.45	0.82	0.61	0.22	0.85
317	0.57	0.42	0.07	—	0.30	—	—	0.49	0.07	0.37
318	2.30	1.70	0.30	—	1.20	—	—	2.00	0.30	1.50
319	1.54	1.54	0.43	0.48	0.94	0.73	1.13	0.97	0.30	1.17
320	1.75	1.74	0.49	0.54	1.07	0.83	1.28	1.09	0.34	1.32
321	0.87	0.37	0.18	0.28	0.61	0.38	0.58	0.38	0.15	0.57
322	0.98	0.41	0.20	0.31	0.68	0.43	0.65	0.42	0.17	0.64
323	0.88	0.35	0.19	0.26	0.66	0.51	0.58	0.36	0.14	0.59
324	1.00	0.40	0.21	0.30	0.75	0.58	0.66	0.41	0.16	0.67
325	0.89	0.36	0.21	0.32	0.63	0.43	0.59	0.37	0.17	0.59
326	1.00	0.41	0.24	0.36	0.71	0.49	0.67	0.42	0.19	0.67
327	0.74	0.38	0.15	0.13	0.49	0.23	0.40	0.34	0.13	0.55
328	0.83	0.43	0.17	0.14	0.55	0.26	0.45	0.38	0.14	0.62
329	0.90	0.36	0.22	0.36	0.64	0.38	0.65	0.39	0.27	0.58
330	1.02	0.41	0.24	0.41	0.72	0.43	0.73	0.44	0.30	0.65
331	0.71	0.31	0.15	0.26	0.47	0.36	0.50	0.32	0.12	0.46
332	0.80	0.35	0.17	0.29	0.53	0.41	0.57	0.35	0.14	0.52
333	0.75	0.30	0.14	0.24	0.48	0.36	0.49	0.31	0.12	0.46
334	0.84	0.34	0.16	0.27	0.54	0.41	0.54	0.34	0.13	0.52
335	1.20	0.57	0.33	0.23	—	0.50	—	0.50	0.21	0.80
336	1.33	0.64	0.37	0.26	—	0.56	—	0.56	0.23	0.89
337	1.35	0.95	0.15	0.13	0.58	0.31	0.49	0.38	0.26	0.57
338	1.54	1.08	0.17	0.15	0.66	0.36	0.56	0.43	0.30	0.65

TABLE 4   Composition of Important Feeds: Amino Acids—*Continued*

Entry Number	Feed Name Description	International Feed Number	Dry Matter (%)	Crude Protein (%)	Arginine (%)	Glycine (%)	Histidine (%)	Isoleucine (%)
	WHEY							
339	dehy (Cattle)	4-01-182	93.0	13.3	0.34	0.49	0.17	0.79
340			100.0	14.2	0.36	0.53	0.18	0.84
341	low lactose, dehy (Dried whey	4-01-186	93.0	16.7	0.60	0.72	0.27	0.96
342	product) (Cattle)		100.0	17.9	0.64	0.77	0.29	1.03
	YEAST *Saccharomyces cerevisiae*							
343	brewers, dehy	7-05-527	93.0	43.8	2.20	1.75	1.09	2.21
344			100.0	46.9	2.35	1.87	1.17	2.37
345	irradiated, dehy	7-05-529	94.0	48.1	2.46	—	1.00	2.94
346			100.0	51.2	2.62	—	1.06	3.13
347	primary, dehy	7-05-533	93.0	48.0	2.60	—	5.60	3.60
348			100.0	51.8	2.81	—	6.05	3.89
	YEAST, TORULA *Torulopsis utilis*							
349	torula, dehy	7-05-534	93.0	49.1	2.64	2.66	1.32	2.85
350			100.0	52.7	2.83	2.85	1.42	3.06

Entry Number	Leucine (%)	Lysine (%)	Methio-nine (%)	Cystine (%)	Phenyl-alanine (%)	Tyrosine (%)	Serine (%)	Threo-nine (%)	Trypto-phan (%)	Valine (%)
339	1.18	0.94	0.19	0.30	0.35	0.25	0.47	0.90	0.18	0.68
340	1.26	1.00	0.20	0.32	0.37	0.26	0.50	0.96	0.19	0.73
341	1.54	1.40	0.41	0.43	0.55	0.46	0.59	0.95	0.27	0.87
342	1.65	1.50	0.43	0.46	0.59	0.49	0.63	1.01	0.29	0.93
343	3.23	3.11	0.74	0.49	1.83	1.50	—	2.12	0.52	2.36
344	3.45	3.33	0.79	0.53	1.96	1.60	—	2.27	0.55	2.52
345	3.56	3.70	1.00	—	2.77	—	—	2.41	0.73	3.06
346	3.79	3.94	1.06	—	2.95	—	—	2.56	0.78	3.26
347	3.70	3.80	1.00	0.50	2.50	—	—	2.50	0.40	3.20
348	4.00	4.10	1.08	0.54	2.70	—	—	2.70	0.43	3.46
349	3.52	3.74	0.77	0.61	2.85	2.00	2.76	2.64	0.52	2.96
350	3.78	4.01	0.83	0.65	3.06	2.14	2.96	2.83	0.56	3.17

TABLE 5   Composition of Important Feeds: Fat and Fatty Acids, Data Expressed As-Fed and Dry (100% Dry Matter)[a]

Entry Number	Feed Name Description	International Feed Number	Dry Matter (%)	Ether Extract (%)	Saturated Fat[b] (%)	Unsaturated Fat[b] (%)	Linoleic Acid (%)	Arachidonic Acid[c] (%)
	ALFALFA *Medicago sativa*							
01	meal dehy 17% protein	1-00-023	92.0	2.3	0.3	0.6	0.40	—
02			100.0	2.5	0.3	0.7	0.43	—
03	leaves, meal dehy	1-00-137	93.0	2.9	0.3	0.8	0.52	—
04			100.0	3.1	0.3	0.9	0.56	—
	ANIMAL							
	tallow—see Fats and oils							
	BARLEY *Hordeum vulgare*							
05	grain	5-00-549	89.0	1.8	0.5	1.3	0.24	—
06			100.0	2.1	0.6	1.4	0.27	—
	COCONUT *Cocos nucifera*							
	oil—see Fats and oils							
	CORN, DENT YELLOW *Zea mays indentata*							
07	grain	4-02-935	89.0	4.0	0.8	3.3	1.82	—
08			100.0	4.5	0.9	3.7	2.05	—
	oil—see Fats and oils							
09	distillers solubles, dehy	5-02-844	93.0	8.8	1.8	7.0	4.46	—
10			100.0	9.5	2.0	7.5	4.80	—
11	gluten, meal	5-02-900	91.0	7.6	1.4	6.2	3.83	—
12			100.0	8.4	1.5	6.8	4.21	—
13	grits by-product	4-03-011	90.0	6.5	1.0	5.5	3.34	—
14	(hominy feed)		100.0	7.2	1.2	6.1	3.71	—
	CRAB *Callinectes sapidus*							
15	process residue, meal (Crab meal)	5-01-663	92.0	1.7	0.5	1.2	0.33	—
16			100.0	1.9	0.5	1.3	0.35	—
	FATS AND OILS							
17	bran oil, rice	4-14-504	100.0	100.0	18.5	81.1	36.50	—
18			100.0	100.0	18.5	81.1	36.50	—
19	fat, swine (Lard)	4-04-790	100.0	100.0	35.9	64.1	18.30	0.3–1.0
20			100.0	100.0	35.9	64.1	18.30	0.3–1.0
21	fat, offal, poultry	4-09-319	100.0	100.0	39.1	60.9	22.30	0.5–1.0
22			100.0	100.0	39.1	60.9	22.30	0.5–1.0
23	oil, coconut	4-09-320	100.0	100.0	90.3	9.7	1.10	—
24			100.0	100.0	90.3	9.7	1.10	—
25	oil, corn	4-07-882	100.0	100.0	12.3	87.7	55.40	—
26			100.0	100.0	12.3	87.7	55.40	—
27	oil, fish, menhaden	7-08-049	100.0	100.0	40.0	60.0	2.70	20.0–25.0
28			100.0	100.0	40.0	60.0	2.70	20.0–25.0
29	oil, flax, common (Linseed oil)	4-14-502	100.0	100.0	8.2	91.8	13.90	—
30			100.0	100.0	8.2	91.8	13.90	—
31	oil, pecan	4-20-525	100.0	100.0	6.9	93.1	30.60	—
32			100.0	100.0	6.9	93.1	30.60	—
33	oil, safflower	4-20-526	100.0	100.0	10.5	89.5	72.70	—
34			100.0	100.0	10.5	89.5	72.70	—
35	tallow, animal	4-08-127	100.0	100.0	47.6	52.4	4.30	0.0–0.2
36			100.0	100.0	47.6	52.4	4.30	0.0–0.2
	FISH							
37	solubles, condensed	5-01-969	51.0	6.5	2.9	3.6	0.20	—
38			100.0	12.8	5.7	7.1	0.39	—
	menhaden, oil—see Fats and oils							
39	menhaden, meal mech extd	5-02-009	92.0	7.7	4.4	3.3	0.11	—
40			100.0	8.4	4.8	3.6	0.12	—
	FLAX *Linum usitatissimum*							
	common, oil (Linseed oil)—see Fats and oils							
41	common, meal solv extd (Linseed meal)	5-02-048	91.0	1.7	0.4	1.3	0.37	—
42			100.0	1.9	0.4	1.5	0.41	—

TABLE 5 Composition of Important Feeds: Fat and Fatty Acids—*Continued*

Entry Number	Feed Name Description	International Feed Number	Dry Matter (%)	Ether Extract (%)	Saturated Fat[b] (%)	Unsaturated Fat[b] (%)	Linoleic Acid (%)	Arachidonic Acid[c] (%)
	MEAT Animal							
43	meal rendered	5-00-385	94.0	9.9	4.70	5.30	0.34	—
44			100.0	10.6	5.00	5.70	0.36	—
45	with blood, meal tankage rendered	5-00-386	92.0	8.1	4.00	4.10	0.28	—
46			100.0	8.8	4.40	4.50	0.30	—
	MILK Cattle							
47	skimmed dehy	5-01-175	94.0	0.9	0.40	0.60	0.01	—
48			100.0	1.0	0.40	0.60	0.01	—
	MILO (SORGHUM GRAIN)—SEE SORGHUM							
	OATS *Avena sativa*							
49	grain	4-03-309	89.0	4.5	1.10	3.50	1.49	—
50			100.0	5.1	1.20	3.90	1.67	—
	PEANUT *Arachis hypogaea*							
51	kernels, meal mech extd (Peanut meal)	5-03-649	92.0	6.7	1.60	5.10	1.25	—
52			100.0	7.3	1.70	5.50	1.36	—
	PECAN *Caya illinoensis*							
	oil—see Fats and oils							
	POULTRY							
53	by-products, meal rendered	5-03-798	93.0	11.6	4.20	7.50	1.72	—
54			100.0	12.5	4.50	8.00	1.98	—
	offal fat—see Fats and oils							
	RICE *Oryza sativa*							
	bran oil—see Fats and oils							
	SAFFLOWER *Carthamus tinctorius*							
	oil—see Fats and oils							
	SKIM MILK—SEE MILK							
55	SORGHUM *Sorghum vulgare*	4-04-383	90.0	2.9	0.60	2.30	1.08	—
56	grain		100.0	3.2	0.70	2.50	1.20	—
57	SOYBEAN *Glycine max*	5-04-610	92.0	1.84	3.00	15.40	7.97	—
58	seeds		100.0	20.0	3.30	16.70	8.66	—
59	flour by-product	4-04-594	90.0	6.0	1.20	4.90	2.96	—
60	(Soybean mill feed)		100.0	6.8	1.30	5.40	3.29	—
61	seeds, meal solv extd	5-04-604	90.0	1.0	0.03	0.07	0.55	—
62			100.0	1.1	0.03	0.08	0.61	—
63	seeds without hulls,	5-04-612	90.0	0.8	0.30	0.50	0.35	—
64	meal solv extd		100.0	0.9	0.30	0.60	0.39	—
	SWINE *Sus scrofa*							
	fats (Lard)—see Fats and oils							
	WHEAT *Triticum* spp							
65	bran	4-05-190	89.0	4.1	0.80	3.30	2.25	—
66			100.0	4.6	0.90	3.70	2.53	—
67	grain	4-05-211	89.0	1.7	0.40	1.30	0.58	—
68			100.0	1.9	0.40	1.50	0.65	—
69	middlings, less than 9.5% fiber	4-05-205	89.0	4.6	0.90	3.70	2.48	—
70			100.0	5.2	1.00	4.10	2.79	—
	WHEY *Bos* spp							
71	dehy (Cattle)	4-01-182	93.0	0.8	0.50	0.30	0.01	—
72			100.0	0.9	0.60	0.30	0.01	—
	YEAST *Saccharomyces cerevisiae*							
73	brewers, dehy	7-05-527	93.0	1.0	0.20	0.70	0.05	—
74			100.0	1.1	0.20	0.80	0.05	—

[a]Data adapted from Edwards (1964), except arachidonic acids values.
[b]Calculated by assuming that ether extract was all triglyceride (except for alfalfa products). Thus, values were calculated by multiplying percent ether extract by fraction which was saturated or unsaturated. Alfalfa ether extract was presumed to be 40% triglyceride equivalent, and the percentage of ether extract was multiplied by 0.04 and then by the fraction which was saturated or unsaturated.
[c]Data adapted from Hilditch and Williams (1964).

TABLE 6 Composition of Important Feeds: Mineral Supplements—*Continued*

Entry Number	Feed Name Description	International Feed Number	Dry Matter (%)	Protein Equivalent N × 6.25 (%)	Calcium (Ca) (%)	Chlorine (Cl) (%)	Magnesium (Mg) (%)
055	carbonate, $CoCO_3$, c p	6-01-565	99.0^b	—	—	—	—
056			100.0	—	—	—	—
057	chloride, hexahydrate, $CoCl_2 \cdot 6H_2O$, c p	6-01-557	98.0^b	—	—	29.20	—
058			100.0	—	—	29.80	—
059	oxide, CoO	6-01-560	99.0^b	—	—	0.01	—
060			100.0	—	—	0.01	—
061	oxide, Co_2O_3, c p	6-01-559	99.0^b	—	—	—	—
062			100.0	—	—	—	—
063	sulfate, heptahydrate, $CoSO_4 \cdot 7H_2O$, c p	6-01-563	100.0	—	—	—	—
064			100.0	—	—	—	—
065	COLLOIDAL CLAY (Soft rock phosphate)	6-03-947	99.0^b	—	17.00	—	0.38
066			100.0	—	17.17	—	0.38
	COPPER (CUPRIC)						
067	carbonate, $CuCO_3Cu(OH)_2 \cdot H_2O$	6-01-703	98.0^b	—	—	—	—
068			100.0	—	—	—	—
069	chloride, dihydrate, $CuCl_2 \cdot 2H_2O$, c p	6-01-706	99.0^b	—	—	41.18	—
070			100.0	—	—	41.60	—
071	gluconate, monohydrate,	6-29-484	99.0^b	—	—	—	—
072	$Cu(C_6H_{11}O_7)_2 \cdot H_2O$, c p		100.0	—	—	—	—
073	hydroxide, $Cu(OH)_2$, c p	6-01-710	98.0^b	—	—	—	—
074			100.0	—	—	—	—
075	orthophosphate, trihydrate,	6-29-485	99.0^b	—	—	—	—
076	$CU_3(PO_4)_2 \cdot 3H_2O$, c p		100.0	—	—	—	—
077	oxide, CuO, c p	6-01-712	99.0^b	—	—	—	—
078			100.0	—	—	—	—
079	sulfate, pentahydrate, $CuSO_4 \cdot 5H_2O$	6-01-719	100.0	—	—	—	—
080			100.0	—	—	—	—
081	sulfate, pentahydrate $CuSO_4 \cdot 5H_2O$, c p	6-01-720	100.0	—	—	—	—
082			100.0	—	—	—	—
	COPPER (CUPROUS)						
083	iodide, CuI, c p	6-01-722	100.0^b	—	—	—	—
084			100.0	—	—	—	—
085	oxide, Cu_2O, c p	6-28-224	99.0^b	—	—	—	—
086			100.0	—	—	—	—
087	CURACAO PHOSPHATE	6-05-586	99.0^b	—	34.00	—	0.80
088			100.0	—	34.34	—	0.81
089	DIIODOSALICYLIC ACID, $C_7H_4I_2O_3$	6-01-787	99.0^b	—	—	—	—
090			100.0	—	—	—	—
091	ETHYLENEDIAMINE DIHYDRIODIDE	6-01-842	98.0^b	—	—	—	—
092	(Organic iodide), $C_2H_8N_2 \cdot 2HI$, c p		100.0	—	—	—	—
	IRON (FERRIC)						
093	ammonium citrate	6-01-857	99.0^b	42.1	—	—	—
094			100.0	42.5	—	—	—
095	chloride, hexahydrate, $FeCl_3 \cdot 6H_2O$, c p	6-28-101	—	—	—	—	—
096			100.0	—	—	39.35	—
097	chloride, $FeCl_3$, c p	6-01-866	98.0^b	—	—	64.27	—
098			100.0	—	—	65.58	—
099	oxide, Fe_2O_3	6-02-431	92.0^b	—	0.30	—	0.40
100			100.0	—	0.32	—	0.43
101	oxide, Fe_2O_3, c p	6-02-432	98.0^b	—	—	—	—
102			100.0	—	—	—	—
103	phosphate, tetrahydrate, $FePO_4 \cdot 4H_2O$, c p	6-29-486	99.0^b	—	—	—	—
104			100.0	—	—	—	—
105	pyrophosphate, nonahydrate,	6-29-487	99.0^b	—	—	—	—
106	$Fe_4(P_2O_7)_3 \cdot 9H_2O$, c p		100.0	—	—	—	—
	IRON (FERROUS)						
107	carbonate, $FeCO_3$	6-01-863	99.0^b	—	—	—	—
108			100.0	—	—	—	—

Entry Number	Phosphorus (P) (%)	Potassium (K) (%)	Sodium (Na) (%)	Sulfur (S) (%)	Cobalt (Co) (%)	Copper (Cu) (%)	Fluorine (F) (%)	Iodine (I) (%)	Iron (Fe) (%)	Manganese (Mn) (%)	Selenium (Se) (%)	Zinc (Zn) (%)
055	—	—	—	—	49.04	—	—	—	—	—	—	—
056	—	—	—	—	49.54	—	—	—	—	—	—	—
057	—	—	—	—	24.27	—	—	—	—	—	—	—
058	—	—	—	—	24.77	—	—	—	—	—	—	—
059	—	—	—	0.20	70.35	—	—	—	0.049	—	—	—
060	—	—	—	0.20	71.06	—	—	—	0.050	—	—	—
061	—	—	—	—	70.35	—	—	—	—	—	—	—
062	—	—	—	—	71.06	—	—	—	—	—	—	—
063	—	—	—	11.40	20.97	—	—	—	—	—	—	—
064	—	—	—	11.40	20.97	—	—	—	—	—	—	—
065	9.00	—	0.10	—	—	—	1.49	—	1.900	—	—	—
066	9.09	—	0.10	—	—	—	1.50	—	1.920	—	—	—
067	—	—	—	0.17	—	55.00	—	—	0.150	—	—	0.02
068	—	—	—	0.17	—	56.12	—	—	0.153	—	—	0.02
069	—	—	—	—	—	36.90	—	—	—	—	—	—
070	—	—	—	—	—	37.27	—	—	—	—	—	—
071	—	—	—	—	—	13.33	—	—	—	—	—	—
072	—	—	—	—	—	13.47	—	—	—	—	—	—
073	—	—	—	—	—	63.83	—	—	—	—	—	—
074	—	—	—	—	—	65.13	—	—	—	—	—	—
075	14.11	—	—	—	—	43.42	—	—	—	—	—	—
076	14.25	—	—	—	—	43.86	—	—	—	—	—	—
077	—	—	—	—	—	79.08	—	—	—	—	—	—
078	—	—	—	—	—	79.88	—	—	—	—	—	—
079	—	—	—	12.80	—	25.40	—	—	—	—	—	—
080	—	—	—	12.80	—	25.40	—	—	—	—	—	—
081	—	—	—	12.84	—	25.45	—	—	—	—	—	—
082	—	—	—	12.84	—	25.45	—	—	—	—	—	—
083	—	—	—	—	—	33.36	—	66.64	—	—	—	—
084	—	—	—	—	—	33.36	—	66.64	—	—	—	—
085	—	—	—	—	—	87.93	—	—	—	—	—	—
086	—	—	—	—	—	88.82	—	—	—	—	—	—
087	14.00	—	0.20	—	—	—	0.54	—	0.350	—	—	—
088	14.14	—	0.20	—	—	—	0.55	—	0.350	—	—	—
089	—	—	—	—	—	—	—	64.44	—	—	—	—
090	—	—	—	—	—	—	—	65.09	—	—	—	—
091	—	—	—	—	—	—	—	78.73	—	—	—	—
092	—	—	—	—	—	—	—	80.34	—	—	—	—
093	—	—	—	—	—	—	—	—	15.840	—	—	—
094	—	—	—	—	—	—	—	—	16.000	—	—	—
095	—	—	—	—	—	—	—	—	—	—	—	—
096	—	—	—	—	—	—	—	—	20.660	—	—	—
097	—	—	—	—	—	—	—	—	33.740	—	—	—
098	—	—	—	—	—	—	—	—	34.430	—	—	—
099	—	—	—	—	—	—	—	—	57.000	0.30	—	—
100	—	—	—	—	—	—	—	—	61.950	0.32	—	—
101	—	—	—	—	—	—	—	—	68.540	—	—	—
102	—	—	—	—	—	—	—	—	69.940	—	—	—
103	13.76	—	—	—	—	—	—	—	24.800	—	—	—
104	13.90	—	—	—	—	—	—	—	25.060	—	—	—
105	20.27	—	—	—	—	—	—	—	24.370	—	—	—
106	20.48	—	—	—	—	—	—	—	24.620	—	—	—
107	—	—	—	—	—	—	—	—	39.600	—	—	—
108	—	—	—	—	—	—	—	—	40.000	—	—	—

TABLE 6 Composition of Important Feeds: Mineral Supplements—*Continued*

Entry Number	Feed Name Description	International Feed Number	Dry Matter (%)	Protein Equivalent N × 6.25 (%)	Calcium (Ca) (%)	Chlorine (Cl) (%)	Magnesium (Mg) (%)
109	fumerate, $FeC_4H_2O_4$	6-08-097	99.0[b]	—	—	—	—
110			100.0	—	—	—	—
111	fumerate, $FeC_4H_2O_4$, c p	6-28-100	99.0[b]	—	—	—	—
112			100.0	—	—	—	—
113	gluconate, $Fe(C_6H_{11}O_7)_2 \cdot 2H_2O$	6-01-867	99.0[b]	—	—	—	—
114			100.0	—	—	—	—
115	oxide, FeO	6-20-728	97.0[b]	—	—	—	—
116			100.0	—	—	—	—
117	oxide, FeO, c p	6-28-108	98.0[b]	—	—	—	—
118			100.0	—	—	—	—
119	sulfate, monohydrate, $FeSO_4 \cdot H_2O$	6-01-869	98.0[b]	—	—	—	—
120			100.0	—	—	—	—
121	sulfate, monohydrate, $FeSO_4 \cdot H_2O$, c p	6-20-725	99.0[b]	—	—	—	—
122			100.0	—	—	—	—
123	sulfate, heptahydrate, $FeSO_4 \cdot 7H_2O$	6-20-734	98.0[b]	—	—	—	—
124			100.0	—	—	—	—
125	sulfate, heptahydrate, $FeSO_4 \cdot 7H_2O$, c p	6-01-870	98.0[b]	—	—	—	—
126			100.0	—	—	—	—
	LIMESTONE						
127	limestone	6-02-632	100.0	—	34.00	0.03	2.06
128			100.0	—	34.00	0.03	2.06
129	magnesium (Dolomitic)	6-02-633	99.0[b]	—	22.08	0.12	9.89
130			100.0	—	22.30	0.12	9.99
	MAGNESIUM						
131	carbonate, $MgCO_3Mg(OH)_2$	6-02-754	98.0[b]	—	0.02	0.00	30.20
132			100.0	—	0.02	0.00	30.81
133	hydroxide, $Mg(OH)_2$, c p	6-29-489	98.0[b]	—	—	—	40.86
134			100.0	—	—	—	41.69
135	oxide, MgO	6-02-756	98.0	—	3.00	—	54.90
136			100.0	—	3.07	—	56.20
137	oxide, MgO, c p	6-02-757	98.0[b]	—	—	—	59.08
138			100.0	—	—	—	60.29
139	sulfate, heptahydrate, $MgSO_4 \cdot 7H_2O$	6-02-758	98.0[b]	—	0.02	—	9.60
140	(Epsom salts)		100.0	—	0.02	—	9.80
141	sulfate, heptahydrate, $MgSO_4 \cdot 7H_2O$, c p	6-02-759	99.0[b]	—	—	—	9.76
142			100.0	—	—	—	9.86
	MANGANESE						
143	dioxide, MnO_2	6-03-042	98.0[b]	—	—	—	—
144			100.0	—	—	—	—
	MANGANESE (MANGANOUS)						
145	acetate, $Mn(C_2H_3O_2)_2$, c p	6-29-490	98.0[b]	—	—	—	—
146			100.0	—	—	—	—
147	carbonate, $MnCO_3$	6-03-036	97.0	—	—	—	—
148			100.0	—	—	—	—
149	chloride, tetrahydrate, $MnCl_2 \cdot 4H_2O$	6-03-038	99.0[b]	—	—	35.47	—
150			100.0	—	—	35.83	—
151	chloride, tetrahydrate, $MnCl_2 \cdot 4H_2O$, c p	6-03-039	99.0[b]	—	—	35.47	—
152			100.0	—	—	35.83	—
153	citrate, $Mn_3(C_6H_5O_7)_2$	6-03-040	99.0[b]	—	—	—	—
154			100.0	—	—	—	—
155	gluconate, $Mn(C_6H_{11}O_7)_2$, c p	6-03-045	99.0[b]	—	—	—	—
156			100.0	—	—	—	—
157	orthophosphate, $Mn(PO_4)_2$, c p	6-29-491	99.0[b]	—	—	—	—
158			100.0	—	—	—	—
159	oxide, MnO, c p	6-03-056	99.0[b]	—	—	—	—
160			100.0	—	—	—	—

Entry Number	Phosphorus (P) (%)	Potassium (K) (%)	Sodium (Na) (%)	Sulfur (S) (%)	Cobalt (Co) (%)	Copper (Cu) (%)	Fluorine (F) (%)	Iodine (I) (%)	Iron (Fe) (%)	Manganese (Mn) (%)	Selenium (Se) (%)	Zinc (Zn) (%)
109	—	—	—	—	—	—	—	—	32.540	—	—	—
110	—	—	—	—	—	—	—	—	32.870	—	—	—
111	—	—	—	—	—	—	—	—	32.540	—	—	—
112	—	—	—	—	—	—	—	—	32.870	—	—	—
113	—	—	—	—	—	—	—	—	11.460	—	—	—
114	—	—	—	—	—	—	—	—	11.580	—	—	—
115	—	—	—	—	—	—	—	—	75.370	—	—	—
116	—	—	—	—	—	—	—	—	77.700	—	—	—
117	—	—	—	—	—	—	—	—	76.180	—	—	—
118	—	—	—	—	—	—	—	—	77.730	—	—	—
119	—	—	—	18.00	—	—	—	—	32.300	—	—	—
120	—	—	—	18.37	—	—	—	—	32.960	—	—	—
121	—	—	—	18.67	—	—	—	—	32.530	—	—	—
122	—	—	—	18.86	—	—	—	—	32.860	—	—	—
123	—	—	—	12.10	—	—	—	—	21.400	—	—	—
124	—	—	—	12.35	—	—	—	—	21.840	—	—	—
125	—	—	—	11.29	—	—	—	—	19.680	—	—	—
126	—	—	—	11.52	—	—	—	—	20.080	—	—	—
127	0.02	0.12	0.06	0.04	—	—	—	—	0.350	—	—	—
128	0.02	0.12	0.06	0.04	—	—	—	—	0.350	—	—	—
129	0.04	0.36	—	—	—	—	—	—	0.076	—	—	—
130	0.04	0.36	—	—	—	—	—	—	0.077	—	—	—
131	—	—	—	—	—	—	—	—	0.021	—	—	—
132	—	—	—	—	—	—	—	—	0.022	—	—	—
133	—	—	—	—	—	—	—	—	—	—	—	—
134	—	—	—	—	—	—	—	—	—	—	—	—
135	—	—	—	—	—	—	0.02	—	—	0.01	—	—
136	—	—	—	—	—	—	0.02	—	—	0.01	—	—
137	—	—	—	—	—	—	—	—	—	—	—	—
138	—	—	—	—	—	—	—	—	—	—	—	—
139	—	—	—	12.75	—	—	—	—	—	—	—	—
140	—	—	—	13.00	—	—	—	—	—	—	—	—
141	—	—	—	12.88	—	—	—	—	—	—	—	—
142	—	—	—	13.01	—	—	—	—	—	—	—	—
143	—	—	—	—	—	—	—	—	—	61.93	—	—
144	—	—	—	—	—	—	—	—	—	63.19	—	—
145	—	—	—	—	—	—	—	—	—	21.96	—	—
146	—	—	—	—	—	—	—	—	—	22.41	—	—
147	—	—	—	—	—	—	—	—	—	46.40	—	—
148	—	—	—	—	—	—	—	—	—	47.80	—	—
149	—	—	—	—	—	—	—	—	—	27.48	—	—
150	—	—	—	—	—	—	—	—	—	27.76	—	—
151	—	—	—	—	—	—	—	—	—	27.48	—	—
152	—	—	—	—	—	—	—	—	—	27.76	—	—
153	—	—	—	—	—	—	—	—	—	30.05	—	—
154	—	—	—	—	—	—	—	—	—	30.36	—	—
155	—	—	—	—	—	—	—	—	—	32.05	—	—
156	—	—	—	—	—	—	—	—	—	32.37	—	—
157	25.05	—	—	—	—	—	—	—	—	22.21	—	—
158	25.30	—	—	—	—	—	—	—	—	22.43	—	—
159	—	—	—	—	—	—	—	—	—	76.67	—	—
160	—	—	—	—	—	—	—	—	—	77.45	—	—

TABLE 6   Composition of Important Feeds: Mineral Supplements—*Continued*

Entry Number	Feed Name Description	International Feed Number	Dry Matter (%)	Protein Equivalent N × 6.25 (%)	Calcium (Ca) (%)	Chlorine (Cl) (%)	Magnesium (Mg) (%)
161	phosphate, trihydrate, $MnHPO_4 \cdot 3H_2O$	6-29-492	99.0[b]	—	—	—	—
162			100.0	—	—	—	—
163	sulfate, monohydrate, $MnSO_4 \cdot H_2O$, c p	6-28-103	100.0	—	—	—	—
164			100.0	—	—	—	—
	ORGANIC IODIDE—see Ethylenediamine dihydriodide						
165	OYSTERSHELL, GROUND (Flour)	6-03-481	99.0	—	37.62	0.01	0.30
166			100.0	—	38.00	0.01	0.30
	PHOSPHATE						
167	defluorinated	6-01-780	100.0	—	32.00	—	0.42
168			100.0	—	32.00	—	0.42
169	rock	6-03-945	100.0	—	35.00	—	0.41
170			100.0	—	35.00	—	0.41
171	rock, low fluorine	6-03-946	100.0	—	36.00	—	—
172			100.0	—	36.00	—	—
173	rock, soft	6-03-947	100.0	—	17.00	—	0.38
174			100.0	—	17.00	—	0.38
	PHOSPHORIC ACID						
175	H_3PO_4	6-03-707	75.0	—	0.04	—	0.38
176			100.0	—	0.05	—	0.51
177	H_3PO_4, c p	6-03-708	85.0	—	—	—	—
178			100.0	—	—	—	—
	POTASSIUM						
179	bicarbonate, $KHCO_3$, c p	6-29-493	99.0[b]	—	—	—	—
180			100.0	—	—	—	—
181	carbonate, K_2CO_3, c p	6-29-495	100.0[b]	—	—	—	—
182			100.0	—	—	—	—
183	chloride, KCl	6-03-755	100.0	—	0.05	47.30	0.34
184			100.0	—	0.05	47.30	0.34
185	chloride, KCl, c p	6-03-756	100.0	—	—	47.55	—
186			100.0	—	—	47.55	—
187	iodate, KIO_3, c p	6-03-758	100.0[b]	—	—	—	—
188			100.0	—	—	—	—
189	iodide, KI	6-03-759	100.0[b]	—	—	—	—
190			100.0	—	—	—	—
191	iodide, KI, c p	6-03-760	100.0[b]	—	—	—	—
192			100.0	—	—	—	—
193	and magnesium sulfate	6-06-177	98.0[b]	—	0.06	1.25	11.60
194			100.0	—	0.06	1.28	11.84
195	sulfate, K_2SO_4	6-06-098	98.0[b]	—	0.15	1.52	0.60
196			100.0	—	0.15	1.55	0.61
197	sulfate, K_2SO_4, c p	6-29-494	100.0[b]	—	—	—	—
198			100.0	—	—	—	—
	SODIUM						
199	bicarbonate, $NaHCO_3$	6-04-272	100.0	—	—	—	—
200			100.0	—	—	—	—
201	bicarbonate, $NaHCO_3$, c p	6-04-273	100.0	—	—	60.66	—
202			100.0	—	—	—	—
203	chloride, NaCl	6-04-152	100.0	—	—	60.66	—
204			100.0	—	—	60.66	—
205	chloride, NaCl, c p	6-20-226	100.0	—	—	60.66	—
206			100.0	—	—	60.66	—
207	fluoride, NaF	6-04-275	100.0[b]	—	—	—	—
208			100.0	—	—	—	—
209	fluoride, NaF, c p	6-04-276	100.0[b]	—	—	—	—
210			100.0	—	—	—	—

Entry Number	Phosphorus (P) (%)	Potassium (K) (%)	Sodium (Na) (%)	Sulfur (S) (%)	Cobalt (Co) (%)	Copper (Cu) (%)	Fluorine (F) (%)	Iodine (I) (%)	Iron (Fe) (%)	Manganese (Mn) (%)	Selenium (Se) (%)	Zinc (Zn) (%)
161	14.97	—	—	—	—	—	—	—	—	62.17	—	—
162	15.11	—	—	—	—	—	—	—	—	62.80	—	—
163	—	—	—	18.97	—	—	—	—	—	32.50	—	—
164	—	—	—	18.97	—	—	—	—	—	32.50	—	—
165	0.07	0.10	0.21	—	—	—	—	—	0.284	0.01	—	—
166	0.07	0.10	0.21	—	—	—	—	—	0.287	0.01	—	—
167	18.00	0.08	4.90	—	0.001	0.002	0.18	—	0.670	0.02	—	0.006
168	18.00	0.08	4.90	—	0.001	0.002	0.18	—	0.670	0.02	—	0.006
169	13.00	0.06	0.03	—	0.001	0.001	3.50	—	1.680	0.02	—	0.01
170	13.00	0.06	0.03	—	0.001	0.001	3.50	—	1.680	0.02	—	0.01
171	14.00	—	—	—	—	—	—	—	—	—	—	—
172	14.00	—	—	—	—	—	—	—	—	—	—	—
173	9.00	—	0.10	—	—	—	1.50	—	1.900	0.10	—	—
174	9.00	—	0.10	—	—	—	1.50	—	1.900	0.10	—	—
175	23.70	0.02	0.03	1.16	0.001	0.001	0.23	—	1.310	0.04	—	0.010
176	31.60	0.02	0.04	1.55	0.001	0.001	0.31	—	1.750	0.05	—	0.013
177	26.86	—	—	—	—	—	—	—	—	—	—	—
178	31.60	—	—	—	—	—	—	—	—	—	—	—
179	—	38.65	—	—	—	—	—	—	—	—	—	—
180	—	39.05	—	—	—	—	—	—	—	—	—	—
181	—	56.58	—	—	—	—	—	—	—	—	—	—
182	—	56.58	—	—	—	—	—	—	—	—	—	—
183	—	50.00	1.00	0.45	—	—	—	—	0.060	—	—	—
184	—	50.00	1.00	0.45	—	—	—	—	0.060	—	—	—
185	—	52.45	—	—	—	—	—	—	—	—	—	—
186	—	52.45	—	—	—	—	—	—	—	—	—	—
187	—	18.27	—	—	—	—	—	59.30	—	—	—	—
188	—	18.27	—	—	—	—	—	59.30	—	—	—	—
189	—	21.00	—	—	—	—	—	68.17	—	—	—	—
190	—	21.00	—	—	—	—	—	68.17	—	—	—	—
191	—	23.55	—	—	—	—	—	76.44	—	—	—	—
192	—	23.55	—	—	—	—	—	76.44	—	—	—	—
193	—	18.50	0.76	22.30	—	—	0.001	—	0.010	0.002	—	0.001
194	—	18.88	0.78	22.76	—	—	0.001	—	0.010	0.002	—	0.001
195	—	41.00	0.09	17.00	—	—	—	—	0.070	0.001	—	—
196	—	41.84	0.09	17.35	—	—	—	—	0.071	0.001	—	—
197	—	44.87	—	18.40	—	—	—	—	—	—	—	—
198	—	44.87	—	18.40	—	—	—	—	—	—	—	—
199	—	—	27.00	—	—	—	—	—	—	—	—	—
200	—	—	27.00	—	—	—	—	—	—	—	—	—
201	—	—	27.36	—	—	—	—	—	—	—	—	—
202	—	—	27.36	—	—	—	—	—	—	—	—	—
203	—	—	39.34	—	—	—	—	—	—	—	—	—
204	—	—	39.34	—	—	—	—	—	—	—	—	—
205	—	—	39.34	—	—	—	—	—	—	—	—	—
206	—	—	39.34	—	—	—	—	—	—	—	—	—
207	—	—	54.75	—	—	—	45.24	—	—	—	—	—
208	—	—	54.75	—	—	—	45.24	—	—	—	—	—
209	—	—	54.75	—	—	—	45.24	—	—	—	—	—
210	—	—	54.75	—	—	—	45.24	—	—	—	—	—

TABLE 6   Composition of Important Feeds: Mineral Supplements—*Continued*

Entry Number	Feed Name Description	International Feed Number	Dry Matter (%)	Protein Equivalent N × 6.25 (%)	Calcium (Ca) (%)	Chlorine (Cl) (%)	Magnesium (Mg) (%)
211	iodate, $NaIO_3$, c p	6-04-278	100.0[b]	—	—	—	—
212			100.0	—	—	—	—
213	iodide, NaI	6-04-279	100.0[b]	—	—	—	—
214			100.0	—	—	—	—
215	iodide, NaI, c p	6-04-280	100.0[b]	—	—	—	—
216			100.0	—	—	—	—
217	phosphate, monobasic, monohydrate,	6-04-288	97.0	—	—	—	—
218	$NaH_2PO_4 \cdot H_2O$		100.0	—	—	—	—
219	phosphate, monobasic, monohydrate	6-04-287	98.0[b]	—	—	—	—
220	$NaH_2PO_4 \cdot H_2O$, c p		100.0	—	—	—	—
221	phosphate, dibasic, from furnace phosphoric	6-04-286	97.0[b]	—	—	—	—
222	acid, Na_2HPO_4		100.0	—	—	—	—
223	selenite, Na_2SeO_3	6-26-013	98.0[b]	—	—	—	—
224			100.0	—	—	—	—
225	selenite, Na_2SeO_3, c p	6-28-104	98.0[b]	—	—	—	—
226			100.0	—	—	—	—
227	selenate, Na_2SeO_4, c p	6-28-105	98.0[b]	—	—	—	—
228			100.0	—	—	—	—
229	sulfate, decahydrate, $Na_2SO_4 \cdot 10H_2O$, c p	6-04-292	97.0[b]	—	—	—	—
230			100.0	—	—	—	—
231	tripolyphosphate, $Na_5P_3O_{10}$	6-08-076	96.0	—	—	—	—
232			100.0	—	—	—	—
	SULFUR						
233	elemental	6-04-705	99.0	—	—	—	—
234			100.0	—	—	—	—
	ZINC						
235	acetate, dihydrate, $Zn(C_2H_3O_2)_2 \cdot 2H_2O$, c p	6-05-548	99.0[b]	—	—	—	—
236			100.0	—	—	—	—
237	carbonate, $ZnCO_3$	6-05-549	99.0	—	—	—	—
238			100.0	—	—	—	—
239	carbonate, tetrahydrate, $5ZnO \cdot 2CO_3 \cdot 4H_2O$	6-29-585	98.0[b]	—	—	—	—
240			100.0	—	—	—	—
241	chloride, $ZnCl_2$, c p	6-05-552	98.0[b]	—	—	51.00	—
242			100.0	—	—	52.03	—
243	oxide, ZnO	6-06-553	100.0	—	—	—	—
244			100.0	—	—	—	—
245	oxide, ZnO, c p	6-05-554	100.0	—	—	—	—
246			100.0	—	—	—	—
247	sulfate, monohydrate, $ZnSO_4 \cdot H_2O$	6-05-555	99.0[b]	—	0.02	0.015	—
248			100.0	—	0.02	0.015	—
249	sulfate, monohydrate, $ZnSO_4 \cdot H_2O$, c p	6-28-106	99.0[b]	—	—	—	—
250			100.0	—	—	—	—
251	sulfate, heptahydrate, $ZnSO_4 \cdot 7H_2O$	6-20-729	98.0[b]	—	—	—	—
252			100.0	—	—	—	—
253	sulfate, heptahydrate, $ZnSO_4 \cdot 7H_2O$, c p	6-05-566	98.0[b]	—	—	—	—
254			100.0	—	—	—	—

[a]The composition of mineral ingredients that are hydrated (e.g., $CaSO_4 \cdot 2H_2O$) is shown including the waters of hydration, both on an as-fed and dry matter basis. Mineral composition of feed grade mineral supplements varies by source, mining site, and manufacturer. Use manufacturer's analysis when available.
[b]Dry matter values have been estimated for these minerals.

Entry Number	Phosphorus (P) (%)	Potassium (K) (%)	Sodium (Na) (%)	Sulfur (S) (%)	Cobalt (Co) (%)	Copper (Cu) (%)	Fluorine (F) (%)	Iodine (I) (%)	Iron (Fe) (%)	Manganese (Mn) (%)	Selenium (Se) (%)	Zinc (Zn) (%)
211	—	—	11.62	—	—	—	—	64.13	—	—	—	—
212	—	—	11.62	—	—	—	—	64.13	—	—	—	—
213	—	—	15.33	—	—	—	—	84.66	—	—	—	—
214	—	—	15.33	—	—	—	—	84.66	—	—	—	—
215	—	—	15.33	—	—	—	—	84.66	—	—	—	—
216	—	—	15.33	—	—	—	—	84.66	—	—	—	—
217	21.80	—	16.18	—	—	—	—	—	—	—	—	—
218	22.50	—	16.68	—	—	—	—	—	—	—	—	—
219	24.88	—	18.47	—	—	—	—	—	—	—	—	—
220	25.39	—	18.85	—	—	—	—	—	—	—	—	—
221	20.85	—	31.04	—	—	—	—	—	—	—	—	—
222	21.60	—	32.00	—	—	—	—	—	—	—	—	—
223	—	—	26.07	—	—	—	—	—	—	—	44.7	—
224	—	—	26.60	—	—	—	—	—	—	—	45.6	—
225	—	—	26.07	—	—	—	—	—	—	—	44.7	—
226	—	—	26.60	—	—	—	—	—	—	—	45.6	—
227	—	—	23.81	—	—	—	—	—	—	—	41.0	—
228	—	—	24.30	—	—	—	—	—	—	—	41.8	—
229	—	—	13.84	9.65	—	—	—	—	—	—	—	—
230	—	—	14.27	9.95	—	—	—	—	—	—	—	—
231	24.00	—	29.80	—	—	—	—	—	0.004	—	—	—
232	25.00	—	31.00	—	—	—	—	—	0.004	—	—	—
233	—	—	—	99.00	—	—	—	—	—	—	—	—
234	—	—	—	99.45	—	—	—	—	—	—	—	—
235	—	—	—	—	—	—	—	—	—	—	—	29.49
236	—	—	—	—	—	—	—	—	—	—	—	29.79
237	—	—	—	—	—	—	—	—	—	—	—	51.63
238	—	—	—	—	—	—	—	—	—	—	—	52.15
239	—	—	—	—	—	—	—	—	—	—	—	53.40
240	—	—	—	—	—	—	—	—	—	—	—	54.50
241	—	—	—	—	—	—	—	—	—	—	—	47.00
242	—	—	—	—	—	—	—	—	—	—	—	47.97
243	—	—	—	—	—	—	—	—	—	—	—	78.00
244	—	—	—	—	—	—	—	—	—	—	—	78.00
245	—	—	—	—	—	—	—	—	—	—	—	80.33
246	—	—	—	—	—	—	—	—	—	—	—	80.33
247	—	—	—	17.50	—	—	—	—	0.001	0.001	—	36.00
248	—	—	—	17.68	—	—	—	—	0.001	0.001	—	36.36
249	—	—	—	17.68	—	—	—	—	—	—	—	36.05
250	—	—	—	17.86	—	—	—	—	—	—	—	36.42
251	—	—	—	10.93	—	—	—	—	—	—	—	22.25
252	—	—	—	11.15	—	—	—	—	—	—	—	22.70
253	—	—	—	10.93	—	—	—	—	—	—	—	22.27
254	—	—	—	11.15	—	—	—	—	—	—	—	22.73

TABLE 7   Stage of Maturity Terms for Plants

Preferred term	Definition	Comparable Terms
For Plants that Bloom		
Germinated	Stage in which the embryo in a seed resumes growth after a dormant period	Sprouted
Early vegetative	Stage at which the plant is vegetative and before the stems elongate	Fresh new growth, before heading out, before inflorescence emergence, immature prebud stage, very immature, young
Late vegetative	Stage at which stems are beginning to elongate to just before blooming; first bud to first flowers	Before bloom, bud stage, budding plants, heading to in-bloom, heads just showing, jointing and boot (grasses), prebloom, preflowering, stems elongated
Early bloom	Stage between initiation of bloom and stage in which $1/10$ of the plants are in bloom; some grass heads are in anthesis	Early anthesis, first flower, headed out in head, up to $1/10$ bloom
Midbloom	Stage in which $1/10$ to $2/3$ of the plants are in bloom; most grass heads are in midanthesis	Bloom, flowering, flowering plants, half bloom, in bloom, midanthesis
Full bloom	Stage in which $2/3$ or more of the plants are in bloom	$3/4$ to full bloom, late anthesis
Late bloom	Stage in which blossoms begin to dry and fall and seeds begin to form	15 days after silking, before milk, in bloom to early pod, late- to postanthesis
Milk stage	Stage in which seeds are well formed but soft and immature	After anthesis, early seed, fruiting, in tassel, late bloom to early seed, past bloom, pod stage, postanthesis, postbloom, seed developing, seed forming, soft, soft immature
Dough stage	Stage in which the seeds are of dough-like consistency	Dough stage, nearly mature, seeds dough, seeds well developed, soft dent
Mature	Stage in which plants are normally harvested for seed	Dent, dough to glazing, fruiting, fruiting plants, in seed, kernels ripe, ripe seed
Postripe	Stage that follows maturity; some seeds cast and plants have begun to weather (applies mostly to range plants)	Late seed, overripe, very mature
Stem cured	Stage in which plants are cured on the stem; seeds have been cast and weathering has taken place (applies mostly to range plants)	Dormant, mature and weathered, seeds cast
Regrowth early vegetative	Stage in which regrowth occurs without flowering activity; vegetative crop aftermath; regrowth in stubble (applies primarily to fall regrowth in temperate climates); early dry season regrowth	Vegetative recovery growth
Regrowth late vegetative	Stage in which stems begin to elongate to just before blooming; first bud to first flowers; regrowth in stubble with stem elongation (applies primarily to fall regrowth in temperate climates)	Recovery growth, stems elongating, jointing and boot (grasses)

TABLE 8   Feed Classes

Class Number	Class Denominations and Explanations
1	**DRY FORAGES AND ROUGHAGES** All forages and roughages cut and cured and other products with more than 18 percent crude fiber or containing more than 35 percent cell wall (dry basis). Forages and roughages are low in net energy per unit weight, usually because of the high cell wall content. *Example forages:* hay; straw; stover (aerial part without ears and without husks (for corn) or aerial part without heads (for sorghum)). *Example roughages:* hulls, pods.
2	**PASTURE, RANGE PLANTS, AND FORAGES FED FRESH** This group comprises all forage feeds either not cut (including feeds cured on the stem) or cut and fed fresh.
3	**SILAGES** This class comprises ensiled forages (corn, alfalfa, grass, etc.), but not ensiled fish, grain, roots, and tubers.
4	**ENERGY FEEDS** Products with less than 20 percent protein and less than 18 percent crude fiber or less than 35 percent cell wall (dry basis), for example, grain, mill byproducts, fruit, nuts, roots, and tubers. When these feeds are ensiled they are classified as energy feeds.
5	**PROTEIN SUPPLEMENTS** Products which contain 20 percent or more protein (dry basis) from animal origin (including ensiled products) as well as oil meals, gluten, etc.
6	**MINERAL SUPPLEMENTS**
7	**VITAMIN SUPPLEMENTS** Including ensiled yeast.
8	**ADDITIVES** Feed supplements such as antibiotics, coloring material, flavors, hormones, and medicants.

TABLE 9   Weight–Unit Conversion Factors

Units Given	Units Wanted	For Conversion Multiply by
lb	g	453.6
lb	kg	0.4536
oz	g	28.35
kg	lb	2.2046
kg	mg	1,000,000.0
kg	g	1,000.0
g	mg	1,000.0
g	μg	1,000,000.0
mg	μg	1,000.0
mg/g	mg/lb	453.6
mg/kg	mg/lb	0.4536
μg/kg	μg/lb	0.4536
Mcal	kcal	1,000.0
kcal/kg	kcal/lb	0.4536
kcal/lb	kcal/kg	2.2046
ppm	μg/g	1.0
ppm	mg/kg	1.0
ppm	mg/lb	0.4536
mg/kg	%	0.0001
ppm	%	0.0001
mg/g	%	0.1
g/kg	%	0.1

TABLE 10   Correlations of Composition with Voluntary Intake by Sheep and with Digestibility[a]

Component	Intake	Digestibility
Digestibility (in vivo)	+0.61	—
Digestibility (in vitro)[b]	+0.47	+0.80
Lignin	−0.08	−0.61
Acid detergent fiber	−0.61	−0.75
Crude protein	+0.56	+0.44
Cellulose	−0.75	−0.56
Cell wall	−0.76	−0.45
Hemicellulose	−0.58	−0.12

[a]For 187 forages of diverse species fed to sheep (Van Soest et al., 1978).
[b]Two-stage procedure of Tilley and Terry (1963).

TABLE 11 Typical Chemical Composition of Crude Protein (CP), Acid Detergent Fiber (ADF), and Neutral Detergent Fiber (NDF) in Alfalfa, Temperate Grasses, and Subtropical Grasses Grown in Florida, Indiana, Pennsylvania, and Wisconsin

Species and Maturity	North			South		
	CP	ADF	NDF	CP	ADF	NDF
ALFALFA						
Bud-first flower	>19	<31	<40	25–30	30–32	33–41
F.F.-midbloom	17–19	13–35	40–46	19–27	34–37	40–47
Mid-full bloom	13–16	36–41	46–51	22	35	42
Postbloom +	<13	>41	>51	17–18	37–41	>51
GRASSES[a]						
Veg-Boot[b]	>18	<33	<55	18–19	32–33	64–69
Boot-early head[c]	13–18	34–38	55–60	8–18	34–40	64–79
Head-milk[d]	8–12	39–41	51–65	6–11	39–43	70–80
Dough +[e]	< 8	>41	>65	4–9	39–47	71–81

[a]North—Smooth bromegrass, orchardgrass, reed canarygrass, and tall fescue.
[b]Subtropical—"Pangola" digitgrass and bermudagrass, 2 to 3 weeks.
[c]Subtropical—Bahiagrass, 2 weeks; "Pangola" digitgrass and bermudagrass, 4 to 6 weeks.
[d]Subtropical—Bahiagrass, 4 to 6 weeks; "Pangola" digitgrass and bermudagrass, 8 weeks.
[e]Subtropical—All 10 weeks.

TABLE 12 Conversion of Beta-Carotene to Vitamin A for Different Animal Species[a]

Species	Conversion of mg of Beta-Carotene to IU of Vitamin A (mg) (IU)	IU of Vitamin A Activity (Calculated from Carotene) (%)
Standard	1 = 1,667	100
Beef cattle	1 = 400	24
Dairy cattle	1 = 400	24
Sheep	1 = 400–500	24–30
Swine	1 = 500	30
Horses		
Growth	1 = 555	33.3
Pregnancy	1 = 333	20
Poultry	1 = 1,667	100
Dogs	1 = 833	50
Rats	1 = 1,667	100
Foxes	1 = 278	16.7
Cat	Carotene not utilized	—
Mink	Carotene not utilized	—
Man	1 = 556	33.3

[a]Taken from Beeson (1965).

REFERENCES

Agricultural Research Council. 1976. The nutrient requirements of farm livestock. No. 4. Composition of British feedstuffs. Agricultural Research Council (obtainable from Her Majesty's Stationery Office, 49 High Holborn, London, W.C.1.).

Armsby, H. 1903. The Principles of Animal Nutrition. 1st ed. John Wiley, New York.

Asplund, J. M., and L. E. Harris. 1969. Metabolizable energy values for nutrient requirements for swine. Feedstuffs 41(14):38–39.

Association of American Feed Control Officials. 1979. Official publication. Ernest A. Epps. Office of the Treasurer, P.O. Box 16390-A., Baton Rouge, LA 70893.

Atwater, W. V. 1874. Annual Report, Connecticut Board of Agriculture.

Beeson, W. M. 1965. Relative potencies of vitamin A and carotene for animals. Fed. Proc. 24:924–926.

Castillo, Leopoldo S., and Amelia L. Gerpacio. 1976. Nutrient composition of some Philippine feedstuffs. Tech. Bull. 21, 3rd ed., Dept. Anim. Sci., University of the Philippines, Los Banos, Philippines.

Crampton, E. W., and L. E. Harris. 1969. Applied Animal Nutrition, 2nd ed. W. H. Freeman, San Francisco.

Crampton, E. W., L. E. Lloyd, and V. G. MacKay. 1957. The calorie value of TDN. J. Anim. Sci. 16:541.

Crampton, E. W., and L. A. Maynard. 1938. The relation of cellulose and lignin content to the nutritive value of animal feeds. J. Nutr. 15:383.

Fonnesbeck, P. V. 1968. Digestion of soluble and fibrous carbohydrate of forage by horses. J. Anim. Sci. 27:1336.

Fonnesbeck, P. V., R. K. Lydman, G. W. Vander Noot, and L. D. Symons. 1967. Digestibility of the proximate nutrients of forage by horses. J. Anim. Sci. 26:1039.

Garrett, W. N. Unpublished data. 1977. Animal Science Department, University of California, Davis.

Goering, H. K., C. H. Gordon, R. W. Hemken, D. R. Waldo, P. J. Van Soest, and L. W. Smith. 1972. Analytical estimates of nitrogen digestibility in heat damaged forages. J. Dairy Sci. 55:1275–1280.

Göhl, B. 1975. Tropical feeds. FAO Feeds Information Center, Animal Production and Health Division, Food and Agriculture Organization of the United Nations, Rome.

Harris, L. E., L. C. Kearl, and P. V. Fonnesbeck. 1972. Use of regression equations in predicting availability of energy and protein. J. Anim. Sci. 35:658.

Harris, L. E., H. Haendler, R. Riviere, and L. Rechaussat. 1980. International feed databank system; an introduction into the system with instructions for describing feeds and recording data.

Publ. 2. Prepared on behalf of INFIC by the International Feedstuffs Institute, Utah State University, Logan.

Harris, L. E., L. C. Kearl, and P. V. Fonnesbeck. 1981. Rationale for naming a feed. Utah Agric. Exp. Stn. Bull. 501.

Henneberg, W., and F. Stohmann (1860, 1864). Beiträge zur Begrundung liner Rationelen Fütterung der Widerkäuer. Vol. 1 and 2. Schwetschke, Brunswick, Germany.

Henry, W. A. 1898. Feeds and Feeding. 1st ed. Published by author, Madison, Wis.

Henry, W. A., and F. B. Morrison. 1910. Feeds and Feeding. 10th ed. Published by authors, Madison, Wis.

Henry, Y. M. 1976. Prediction of energy values of feeds for swine from fiber content. First International Symposium on Feed Composition, Animal Nutrient Requirements, and Computerization of Diets, Utah State University, Logan.

Jones, D. B. 1941. Factors for converting percentages of nitrogen in foods and feeds into percentages of protein. USDA Circular 183.

Kearl, L. C., M. Farid, L. E. Harris, M. Wardeh, and H. Lloyd. 1978. Middle East Feed Composition Tables. International Feedstuffs Institute, Utah State University, Logan.

Knight, A. D., and L. E. Harris. 1966. Digestible protein estimation for NRC feed composition tables. Proc. West. Sec., Am. Soc. Anim. Sci. 17:283, and J. Anim. Sci. 25:593.

McDowell, L. R., J. H. Conrad, J. E. Thomas, and L. E. Harris. 1974. Latin American Tables of Feed Composition. Department of Animal Science, University of Florida, Gainesville.

Moe, P. W., and H. F. Tyrrell. 1976. Estimating metabolizable and net energy of feeds. First International Symposium Feed Composition, Animal Nutrient Requirements, and Computerization of Diets, Utah State University, Logan.

Morrison, F. B., and Associates. 1936. Feeds and Feeding. 20th ed. Morrison Publishing, Ithaca, N.Y.

National Research Council. 1956. Composition of Concentrate By-Product Feedingstuffs. Publ. 449. National Academy of Sciences, Washington, D.C.

National Research Council. 1958. Composition of Cereal Grains and Forages. Publ. 585. National Academy of Sciences, Washington, D.C.

National Research Council. 1966. Biological Energy Interrelationships and Glossary of Energy Terms. Publ. 1411. National Academy of Sciences, Washington, D.C.

National Research Council. 1971. Atlas of Nutritional Data on United States and Canadian Feeds. National Academy of Sciences, Washington, D.C.

National Research Council. 1981. Nutritional Energetics of Domestic

Animals and Glossary of Energy Terms. National Academy Press, Washington, D.C.

Pichard, G., and P. J. Van Soest. 1977. Protein solubility of ruminant feeds. Proc. Cornell Nutr. Conf., Ithaca, N.Y. pp. 91–98.

Preston, R. L. 1972. Protein requirement for growing and lactating ruminants. *In* Sixth Nutrition Conference for Feed Manufacturers, pp. 22–37. H. Swan and D. Lewis, eds. Churchill Livingstone, Edinburgh and London.

Rohweder, D. A., R. F. Barnes, and N. Jorgensen. 1978. Proposed hay grading standards based on laboratory analyses for evaluating quality. J. Anim. Sci. 47:747–751.

Sibbald, I. R. 1977. The true metabolizable energy values of some feedingstuffs. Poult. Sci. 56:380.

Swift, R. W. 1957. The caloric value of TDN. J. Anim. Sci. 16:1055.

Tilley, J. M. A., and R. A. Terry. 1963. A two-stage technique for the *in vitro* digestion of forage crops. J. Br. Grassl. Soc. 18:104–111.

Tyler, C. 1975. Albrecht Taher's hay equivalent: Fact or fiction? Nutrition Abstr. Rev. 45(1):1–11.

Van Soest, P. J., D. R. Mertens, and B. Dienum. 1978. Preharvest factors influencing quality of conserved forage. J. Anim. Sci. 47:712–720.

Van Soest, P. J., and R. H. Wine. 1968. Method for determination of lignin, cellulose and silica. J. Assoc. Off. Anal. Chem. 51:780.

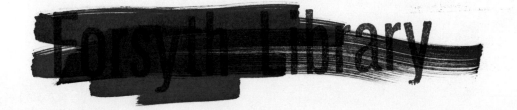